Human Factors in
Process Plant Operation

Human Factors in Process Plant Operation

DAVID A. STROBHAR, PE

MP MOMENTUM PRESS

MOMENTUM PRESS, LLC, NEW YORK

Human Factors in Process Plant Operation
Copyright © Momentum Press®, LLC, 2013.

First published in 2012 by Momentum Press®, LLC
222 East 46th Street, New York, NY 10017
www.momentumpress.net

ISBN-13: 978-1-60650-463-5 (hard cover, case bound)
ISBN-10: 1-60650-463-0 (hard cover, case bound)
ISBN-13: 978-1-60650-465-9 (e-book)
ISBN-10: 1-60650-465-7 (e-book)

DOI: 10.5643/9781606504659

Cover Design by Jonathan Pennell
Interior design by Exeter Premedia Services Private Ltd.,
Chennai, India

10 9 8 7 6 5 4 3 2 1

Printed in the United States of America.

To Andrew David Strobhar (1987–2004)

My son lives in the heart of heaven,
and I live in the heart of earth,
but we live together in the heart of God.

—Ann Weems

Contents

Foreword

Human Factors in Process Plant Operation by David A. Strobhar is a very important book. Forewords often start with wording like that, but in this case it is true and very sincerely meant. It took me over two years to convince Dave Strobhar to write it, and I am deeply glad and impressed that he did. Well done! And I am very proud to have been asked to write the Foreword. I hope you read it closely, because this book can literally save lives.

It is only recently that Human Factors Engineering has been applied to the control of process plants in a systematic way, and only a little less recently that Human Factors has been applied at all. Plant control systems began with open control loops with humans watching dials and turning valves as needed. They moved to closed loop control, digital control and distributed control without thinking about how operators would use the tools or see the necessary view. We went from walls of panel mounted displays to tiny CRT screens with no thought that this might impact operations and safety, and we were, as Dave Strobhar shows, wrong.

The issues in process plants that are affected by the design of control systems are management of alarms, optimization of plant production and throughput, and safety and security of the plant and its workers.

This book takes on all of those issues and shows how they are interrelated, and how, left uncorrected, poor alarm management can lead very easily to deaths and injuries; how it isn't possible for operators with bad system overviews to optimize the plant; and how process safety and industrial control system security are affected by bad, or even poor design of control systems.

Dave Strobhar's book explains how human factors engineering concepts must be used as the basis of control systems design and operation. The book lays out the principles of human ergonomics and how they must be incorporated into any control system from the very beginning in order to insure safe and efficient operations.

Controls engineers, system integrators, operators and managers will all gain incomparable, inside-the-controls industry experience from Strobhar's clear discussion of performance-shaping factors.

Strobhar does something unique by describing graphically controls in terms of workload and staffing, along with a detailed analysis of the all-important human-controls interface, including workspace, content and format. This insight alone is worth the price of the book. But as the infomercials say, "Wait, there's more!"

Strobhar discusses how-to planning for system demands and levels of automation and provides unique guidance on worker selection and training, along with example procedures and job aids;.

Finally, Strobhar analyzes the 2005 Texas City incident and shows what happened, why it happened, and how it could have been avoided and prevented.

There are literally hundreds of accidents annually in the process industries worldwide that produce injuries, fatalities, and loss of production. The Abnormal Situation Consortium has an RSS feed with daily accounts of such incidents. A significant number of those accidents have as their root cause an error by an operator or a maintenance problem that was not corrected. As you will see in Dave's analysis of the 2005 BP Texas City disaster, human factors, and poor human factors engineering, contributed greatly to the loss of life and the enormous loss of property and production.

It seems like every time a significant accident occurs, there's a flutter of activity toward making plants safer for operators and techs to work in, but then it stops. Dr. John Lambshead, professor emeritus at the London Museum of Natural History, calls this the "leopard in the grass" reaction. He describes it. "Long ago, when barely human hominids roamed the savannah, the grass rustled and a leopard leaped out and the unlucky one was leopard lunch. The rest of the band ran away. For a while after that, every time the grass rustled, everybody ran, but soon it was noticed that there was no leopard. So complacency sets in, and nobody runs...until another leopard leaps out."

So we have managed to plateau our response to safety issues, because we don't continue to be watchful. But as Dave Strobhar points out, we can design systems from the ground up to be watchful for us, so that we don't have to worry about losing situational awareness. Our systems can maintain situational awareness for us. This is how we will get past the leopard.

Walt Boyes, Life Fellow, ISA; Fellow InstMC
Chartered Measurement and Control Technologist
Spitzer and Boyes LLC

Acknowledgments

This book is the product of over 30 years of work at Beville Engineering. As such, I am indebted to all who have worked with me over the years. This book would not have been possible without them.

Two of my Beville associates deserve special recognition and thanks. Lisa Via and Kim Elkins have been working alongside me for many years and are instrumental to the success of both Beville Engineering and the contents of this book. I could not have done either without them. Their work, support, sacrifice, insights, and dedication are irreplaceable.

I want to thank individuals at several plants. Ted Coonan of Shell was both a mentor and maintained faith in what human factors and what I had to offer through projects both good and bad. Pat Kimmet and John Berg of Cenex provided support during the most difficult time of my life.

Both Gary Broughton and Chris Christiansen, my first supervisors, gave me invaluable career guidance.

Finally, this book would not have occurred without the encouragement of the members of the Center for Operator Performance.

Notice: Figures appear in color in the E-book edition of this book.

1

Introduction

> I was asked by an engineer at a major chemical company, "Can you look over this list of tasks we developed? We are trying to incorporate human reliability into our hazard analysis." Upon looking at the list, which was a list of things (e.g., "compressor controls" and "charge pump"), I pointed out, "There are no verbs in this list, so these are not tasks." The engineer was puzzled, seeming to not understand my point. His qualification to analyze human performance appeared to be primarily related to his being human. Unfortunately, that is rarely enough.

I graduated from Wright State University in 1980 with a degree in human factors engineering. At that time, only a few people knew of either human factors engineering or its European counterpart, ergonomics. My adult children still aren't quite certain about what it is that I do. When my son was asked in elementary school what his father was doing, he said, "My dad tells people smart things." I ran into an old friend a few years ago, and he asked, "Are you still telling people that red means stop?" (Ironically, it usually means "warning" in process plants. I am telling people it shouldn't also mean stop.) Despite the lack of understanding, or perhaps because of it, the terms *human factors* and *ergonomics* have become more commonplace. Even a mattress manufacturer describes a mattress as "ergonomic," which is odd, given that ergonomics is the study of work (*ergo* is Greek for work). I have worked to go to sleep but rarely considered it work.

Although human factors begins with the word human and is about humans, the need to consider human factors is the product of technology. It was the massive technological advancements of the Second World War that highlighted the problem of ignoring the human component of the system [1].

As large numbers of soldiers were needed, while using increasingly sophisticated systems, the rate of "operator error" dramatically increased. The Army Air Corp in the Second World War saw aircraft being built that many people could not fly. Had we reached a limit of human performance, that is, people can only control vehicles below half the speed of sound? No. A lesson to be learned is that we humans are capable of doing amazing things, given the right tools. If you look at the two graphs in Figure 1.1, the problem becomes easy to see (Julian Christensen, pers. comm.). As the speed and capability of the aircraft was beginning to exponentially increase, so too was the number of controls and switches with which the pilot had to contend. The pilots had the problem of information overload.

A study was carried out by a group of psychologists led by Dr. Paul Fitts. They began to look at how to design the system for the user rather than making the user adapt to the system. A classic study was to identify the key information that a pilot utilize to fly the plane and format it into a standard configuration in the cockpit [2]. They identified what became known as the sacred six, the six basic pieces of information that a pilot need to fly a plane. All the rest of the displays were necessary for some aspect of the aircraft or mission, but these six instruments (airspeed, gyro, horizon, altimeter, turn/ bank, and rate of climb) are what is needed to a plane fly. If a pilot ran into a problem, they should focus on these instruments. They should keep the plane in the air, complete the mission if possible, or return to base if necessary, but keep the plane and pilot intact to fly again. They should separate the truly critical information from the less critical.

As technology has become more pervasive, there is a need to consider how a person can use it. My first job out of college was at GPU Nuclear Corp. We operated three nuclear power plants, with different capacities and alarms as shown in Figure 1.2: The 1968 vintage plant had a capacity of 680 MW with 300 hardwired alarms; the 1974 vintage plant had a capacity of 820 MW with 600 hardwired alarms; and the 1978 vintage plant had a capacity of 820 MW and a sister plant to the one built 4 years earlier, with 1200 hardwired alarms. The last plant was Three Mile Island Unit 2, where an incident occurred in 1979 due to an "operator error," which ended the construction of nuclear power plants in the United States for decades. As shown in Figure 1.2, the alarms were increasing exponentially, actually faster than the increase in plant output. One of the contributing causes of the accident was alarm overload. The same sort of overload problem the U.S. Air Force had encountered three decades before had struck the U.S. nuclear power industry.

What is the major technological change in process plants of the last generation? I would argue that it is the ubiquitous nature of distributed control systems (DCS). This provided the accelerant to human factors issues in refineries and chemical plants, as the DCS enabled the addition of alarms with

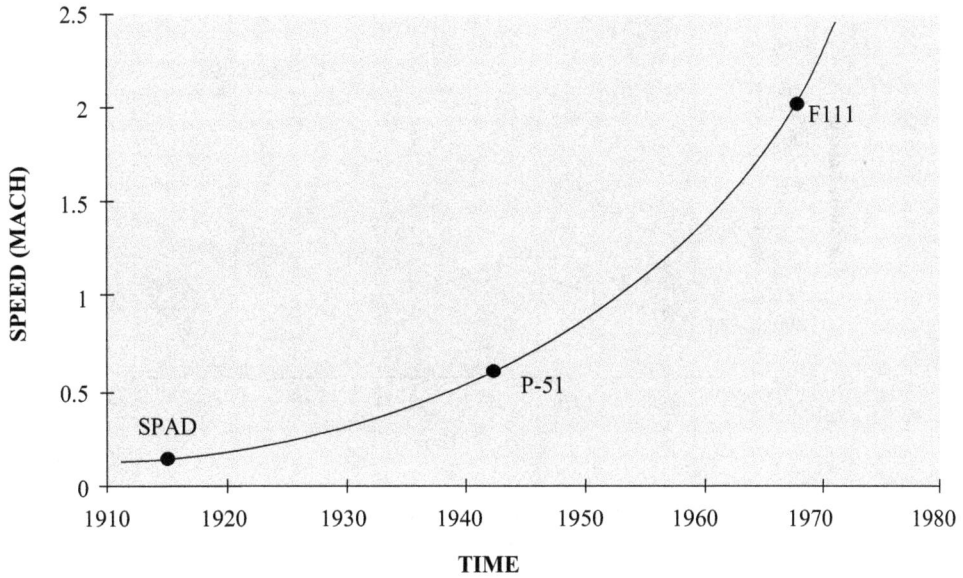

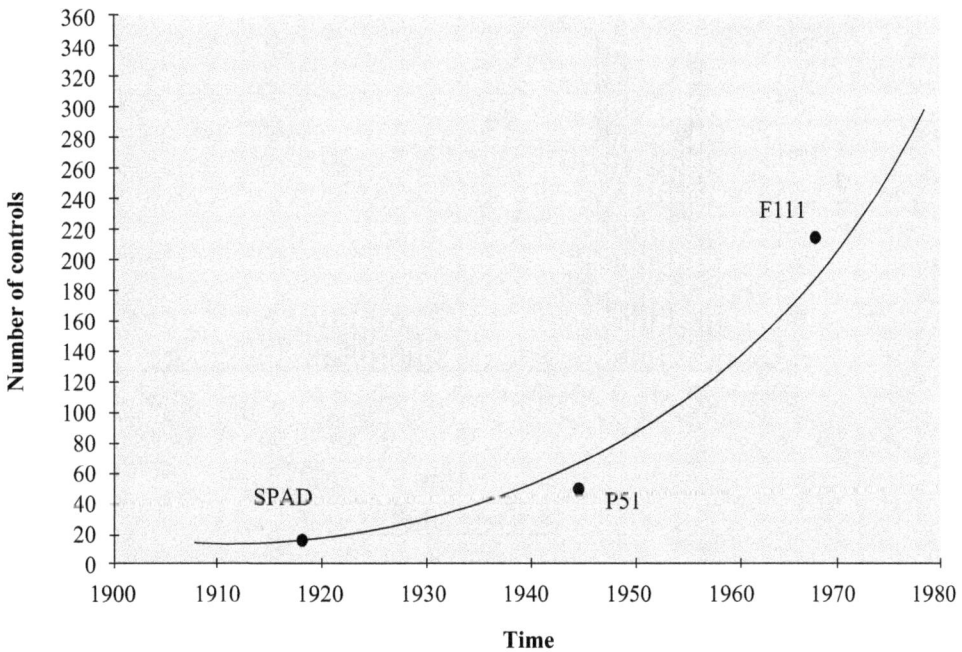

Figure 1.1. Aircraft Speed and Instrumentation.

little cost to anyone but the operator trying to deal with them. The results of easily adding alarms were situations similar to that seen in Figure 1.3. This is a 200,000 barrel per day crude oil unit. Each bar is a 5 min increment of time. The first 15 min is the beginning of the shift, wherein the operator is getting

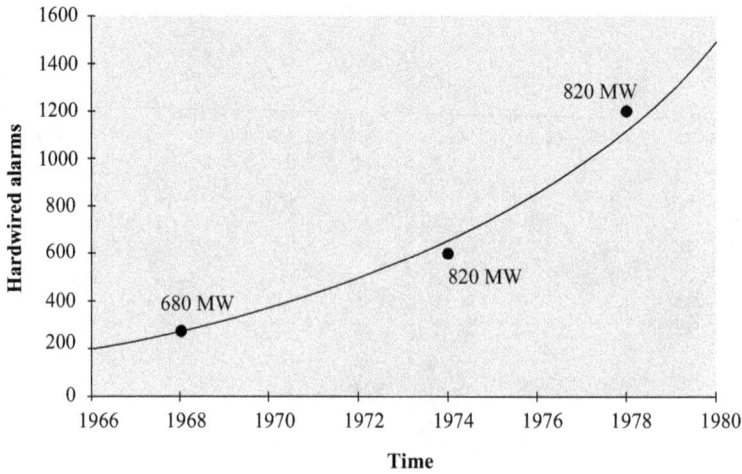

Figure 1.2.   Nuclear Power Plant Alarms Over Time

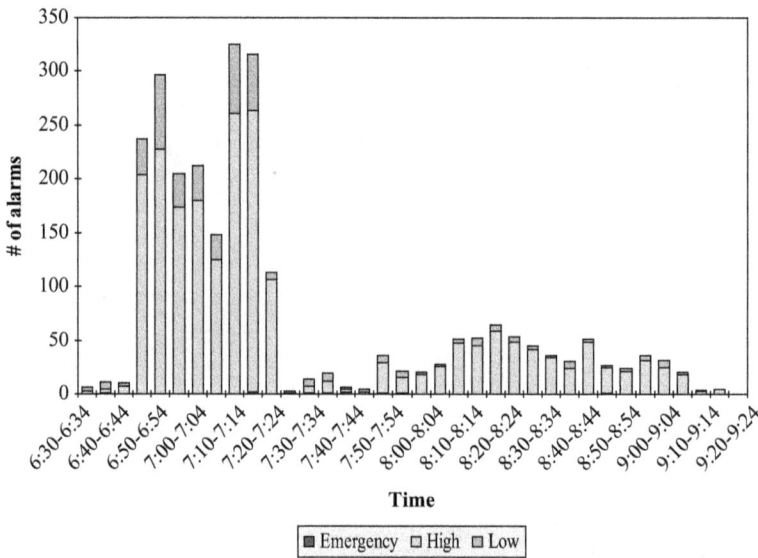

Figure 1.3.   Histogram of Crude Alarm Actuations.

a few alarms every 5 min. At about 20 min into the shift, the operator turns on the advanced control system on the fired heaters, incorrectly. The result is over 200 alarms per 5 min for the next 20 min. He then gets a 15 min break of only 150 alarms per 5 min before being slammed with over 300 alarms per 5 min for the next 10 min. That rate of alarm actuation is both above the rate for comprehension and even for signal detection (this will be discussed in detail later).

The operator in this case quickly abandoned the alarm summary screen provided on the system. He realized he could not keep up with the alarm rate and resorted to paging through key graphics. The unit did not shut down, although he did send off-spec material to a downstream unit and poisoned their catalyst. While this operator got through the upset reasonably well, the same cannot be said for another unit at the same refinery. A few months later, the refinery suffered a power failure, creating the same sort of alarm flood on all the consoles, including that for a sulfur recovery unit. That operator used the same strategy of monitoring key graphics but failed to observe a loss of seal leg and the release of deadly hydrogen sulfide gas into the unit. A tragedy was averted only because the field operator had chosen to put on fresh air equipment at the beginning of the upset. Perhaps the operator realized it was a severe upset. Perhaps the operator was just cautious. Perhaps he knew who was on the board that night. When the operators needed the alarm system the most, it was of the least value to them.

Since that time, human factors has become a more common, if not often misunderstood, term in process plant operation. The publication of OSHA 1910.119 (Process Safety Management of Highly Hazardous Chemicals) in the early 1990s specified that human factors were "to be considered." Of course, no one knew what that meant and anyone could claim to have "considered it." Problems with the alarm systems resulted in the publication of EEMUA 191 (Alarm Systems: A Guide to Design, Management, and Procurement) in 1999 and the U.S. equivalent in the form of ISA18.2 (Management of Alarm Systems for the Process Industries) in 2009. More recently, the PHMSA (Pipeline Hazardous Materials Safety Administration 49 CFR 195.446) and ISO (International Standards Organization 11064: Ergonomic Design of Control Centres) control room standards have highlighted the need to address human factors in the design and operation of complex systems.

Despite the increase in popularity, many in the industry fail to grasp the whole range of issues that are human factors. They think they know human factors by virtue of their being human. This book is intended to be a primer on some of the key human factor issues and concepts in process plant operation. I have found that one of the problems in the industry is that human factors engineering is made to be overly complicated or overly simplistic in its application. Often, this is a product of a simple failure to realize the basic approach to design the man–machine systems combined with a naivete regarding what enables us to function in this increasingly complex world. This book can help address both.

2

Human Information Processing

In 1992, an oil company requested us to develop a display standard. The company needed us to create displays in a novel way for its distributed control system. We asked the company's personnel as to how they currently would go about it. They said, "We lay out the piping and instrumentation drawings. Using an 8.5×11" sheet of paper, we sketch rectangles on all the drawings until all of the instrumentation has been captured." They then sent the drawings to be configured with a display to match each rectangle.

They at least suspected that their approach was less than optimal. They lamented that courses in display design by the distributed control systems (DCS) vendor only taught them skills to assemble parts of the DCS but did not really explain what the system should look like when the design is completed. They felt, "It's like learning how to work with brick and steel, and being told to build a bridge, with no idea as to what the bridge should look like upon completion."

Good engineering requires an understanding of the characteristics of the components that are used in the design. Chemical engineers should be aware of the boiling points and reactivity of various components in a process. Mechanical engineers should know about metallurgy of the piping or the heat transfer characteristics of an exchanger. Electrical engineers should understand resistances and electrical loads. Good human factors engineering requires an understanding of how humans process information.

A common model of operator in the loop performance is shown in Figure 2.1 [3]. A process undergoes a disturbance $w(t)$, resulting in a change

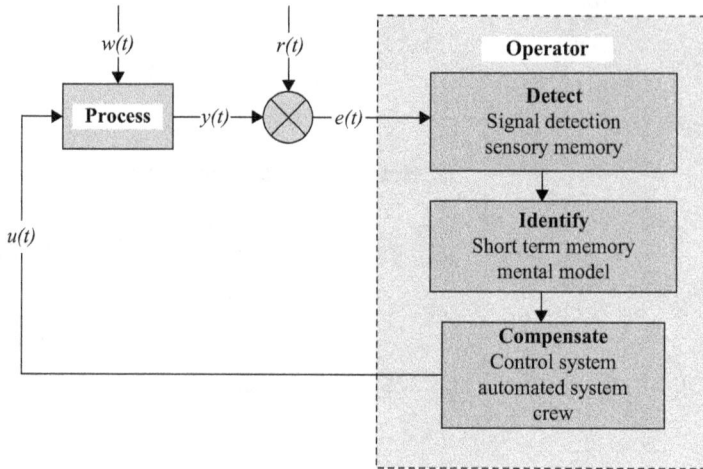

Figure 2.1.   Model of Operator in the Loop Performance.

in output $y(t)$. When the output varies enough from its reference point $r(t)$, the operator detects the change $e(t)$, identifies the cause, and compensates for it, effecting a change on the process $u(t)$. Before discussing the factors responsible for this change, it would be useful to view the larger picture of this model. What is important is the output of what is called the operator subsystem, that is, the change(s) made to bring the process back to the desired reference point. All the inputs to the operator are relevant only to the extent they help to affect a change. What you "know" becomes less important. What is relevant is what you need to do. Information that does not assist in generating the output can get in the way of the process, as will be evident in the following discussion.

Improper output from the operator subsystem, either the wrong or no output, is what is typically called an operator error. These can either be (i) not doing something that should be done (errors of omission) or (ii) doing something that shouldn't be done (errors of commission). While failures can occur at all three steps (detection, identification, compensation), most of them in process plants occur due to failures in either detection or identification. Either the operators didn't realize the problem, or they incorrectly identified the cause and applied a corresponding solution to it. While there have been some instances of an operator knowing what needs to be done improperly applying the solution (such as turning the wrong valve), such incidents are rare.

There are features of the human information processing system that impact our ability to perform each stage in the model. Our ability to detect the problem is influenced by our basic ability to detect signals and use our sensory memory system. Identifying the cause of the disturbance is influenced by characteristics of our short-term memory (conscious) and our mental model of the process. And finally, to effectively compensate for the disturbance, the number of actions that must be taken and the degree to which they are automated determine our response.

2.1 SIGNAL DETECTION

As shown in the model, if the operator fails to detect the error signal, $e(t)$, no output will be forthcoming from the operator. Depending upon the disturbance, the operator may only get a report in their file. So detection of a problem is the first step and its failure is repeatedly seen in incident investigations. An operator at one plant went around 20 min without noticing lack of flow through one pass of a multi-pass fired heater after a control valve failed. When the lack of flow was detected, the valve was thrown wide open, which promptly caused the tube to split and the heater to burn down. The operator took no action for 20 min because they failed to detect that the process had deviated from normal.

Failure to detect a signal is generally because (i) the rates at which the signal is being presented are too high, (ii) the signal fails to look significantly different from the visual noise in the system, or (iii) the operator has tuned out. Signals can come too fast for us to detect them, like the candy on the assembly line in the "I Love Lucy Show." However, that rate is far higher than many people would guess, something like 25 events per minute [4]. Although this number is high, that is only half of the alarm rate that some operators have experienced. Clearly, at alarm rates of 25 alarms per minute or greater, the threshold for human signal detection is exceeded. To say that an operator "missed" an alarm at those rates is to really say that he was performing as expected. He was being human.

Assuming a less insane rate of five signals per minute, the operator can still fail to detect them. This is usually because the signals are lost as they don't stand out. In any visual system, there is both the event the operator has to notice (the signal) and everything else (the noise). If the signal and noise have very similar characteristics, it becomes very hard to detect the signal. Consider the task of detecting the letter "C" as the signal in Figure 2.2. As the noise increases and/or the signals become more like the noise, the harder it is to detect the signal.

To ensure that the signal is detected, either the amount of visual noise has to be reduced and/or the salience of the signal increased. Reducing visual noise is achieved by eliminating visual elements that do not contribute to the operator taking the correct action, such as static information. Extraneous lines and detail add to visual noise. Increasing the strength of the signal is done by

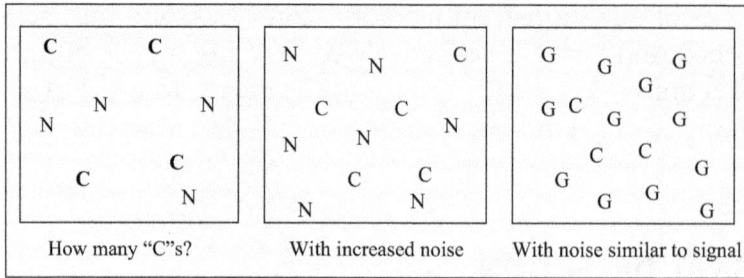

Figure 2.2.   Signal with Varying Levels of Noise.

making it stand out with a unique color, size, or shape. I was at a plant with a new distributed control system in a new control room. Mounted on top of the monitors and wrapping the console were 14 annunciator panels with a 4 × 8 array of alarm tiles. All of the tiles were red. When I inquired as to why they were all red, the operator told me, "Because they are important alarms." However, if they are all red, then the most critical are not going to have any greater signal strength than the less critical. If everything is critical, nothing is critical. The design of any piece of visual information should be examined to determine the amount of visual noise present and whether the signal can be detected.

The final element in our ability to detect signals is our desire to do so, that is, our motivation. Some of the initial research in this area was done on sailors attempting to spot incoming torpedoes from enemy ships, a task with a high motivation factor. One of the major (if not *the* major) benefits of an alarm system is to alert the operator that something needs to be detected, that is, to change their motivation. Cry-wolf alarms have the opposite effect, lowering the operator's motivation to detect a problem by lowering the probability that the alarm really is a signal.

2.2   SENSORY MEMORY

Aiding in the detection of events is a unique feature of our information processing system, which is the sensory memory. Even before we consciously process information, our minds are assigning attributes to the stream of data it receives. Colors and patterns are continuously processed by the sensory memory system. This preprocessing can be seen when it comes into conflict with conscious processing. A psychological test called the Stroop Word-Color test is available on the Internet, in which you try to quickly say out loud the *color* of words like those in Figure 2.3 [4]. The preprocessing of sensory memory

| RED | BLUE | Green | Yellow |

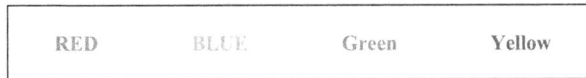

Figure 2.3.   Stroop Word-Color Test.

conflicts with the mental processing of the words, resulting in errors (and a few laughs). However, what this means to the designer is that color and patterns can be processed even if the conscious processing system is full or overloaded.

There is one caveat to utilizing this preprocessing—there can be only a single attribute for the color or pattern. If a color has two possible meanings, for example red meaning both "off" and "warning," then it must be consciously processed. The "meaning" of the color becomes dependent on its use and the context of the display and in this case the preprocessing advantage disappears. This is the basis for the utilization of colors for a single specific use. And it forms the basis for the guideline in the creation of any code: The coding element must have a unique meaning associated with it.

2.3   SHORT-TERM MEMORY

Once a signal has been detected, then it is subject to conscious processing, which happens in short-term memory (STM). Also called working memory, STM is a capacity-limited system in that it can only hold only a limited amount of information, which is about 7 ± 2 chunks of information [5]. Think of STM as a set of receptacles for information, like in Figure 2.4. Chunks of information to be processed fill each receptacle and are discarded when no longer needed. However, if more than seven chunks show up at the same time, then one of two things must happen. Either new information must be ignored, or older information must be jettisoned. This is generally the source of "information overload"; if you have more than seven chunks in a short time, then something is not going to be consciously processed.

Seven, while a lucky number, is not a lot. How can we function in a world where we are bombarded with billions of bits of information? The key is that the STM handles chunks of information, not bits. So the trick is to enable as many bits of information as possible to be handled as a single chunk. This is the rationale behind phone numbers and social security numbers having hyphens. It breaks the string into chunks. The aphorism that a picture is worth a thousand words captures the essence in overcoming the limits of short-term memory.

In a strange twist, older control room technology often created better chunking of alarm information than the first generation control systems did. While likely not developed because they enable better chunking of information,

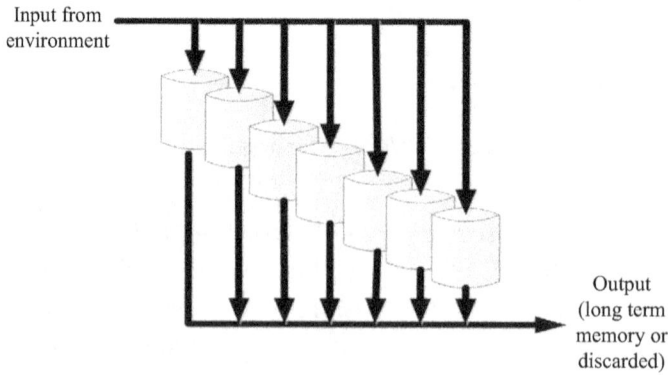

Figure 2.4.   Short Term Memory (STM) Information Processing.

Figure 2.5.   Hardwired Annunciator Panels. *Source*: ©iStock.com/Buretsu

hardwired annunciator panels like those in Figure 2.5 could allow the operator to process a large amount of data. At least when first built, the alarms in a given box were generally related to and associated with the controls below them. If a major upset occurred, the operator could scan the control room to determine which boxes had alarms and which didn't. The operator was not processing each tile or alarm, but the box or system as a whole to understand the problems and their magnitude. So in a control room with 12 annunciator panels of 4 × 8 arrays of alarms, the operator would process at one level just 12 boxes (or sometimes less as multiple light boxes were associated with major sections of the plant), rather than the 384 individual tiles/alarms.

Of course, this approch only works as long as all the alarms in a given box are related to the system breakdown. Modifications to the plant after it was built generated more alarms than there were spaces in the light box. Instead, an open spot was found, often with alarms having nothing to do with the modification. This destroyed the ability to process the light boxes as chunks. Now each alarm had to be processed individually, rather than as part of a larger chunk.

Contrast the ability to process almost 400 alarms as 12 chunks of information versus the early alarm displays for distributed control systems. Alarms were typically presented as a list by time of actuation, showing the time, tag, descriptor, and perhaps the current and/or alarm value, similar to that in Figure 2.6. The sequence of alarms would vary even for the same event, and that sequence could have alarms from all across the operator's span of control. Each alarm had to be processed individually, as no opportunity existed to mentally group them. How many alarms could an operator process in a short period of time with this type of presentation? If you said seven, plus or minus two, you would likely be pretty close to the mark. Hopefully, you already see how the presentation of the data can affect its ability to be processed.

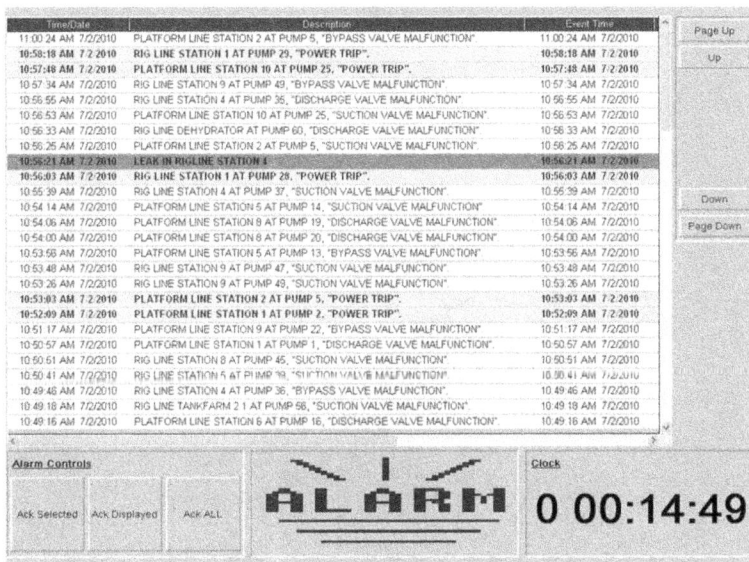

Figure 2.6. Chronological Alarm Display.

By aforementioned discussion, I am not arguing for a return of hardwired annunciator panels. The lesson is not that light boxes are better, but that there was a characteristic of the light boxes that enabled a higher rate of information processing than did the alarm summary display. The goal is to embody that same characteristic in the display system, but with greater power and flexibility using modern control system technology.

More importantly, it is not a stipulation that a display or procedure should only have seven items on it. The goal should be to provide as much data as possible, organized and structured such that the data can be processed as chunks. With large systems, a good designer will be making chunks of chunks, similar to that in Figure 2.7. Data are organized to create a chunk, and that chunk is grouped with other chunks to form a higher-level chunk. This is how we humans function in a world in which we must process billions of bits of data per minute.

An easy technique for chunking information is by enclosing the related information with demarcation lines. The related information can be put inside a box. Consider the two displays in Figure 2.8. Both displays have exactly the same information. The difference is the use of demarcation lines in the first. The display on the left is broken first into two parts, one for each fired heater.

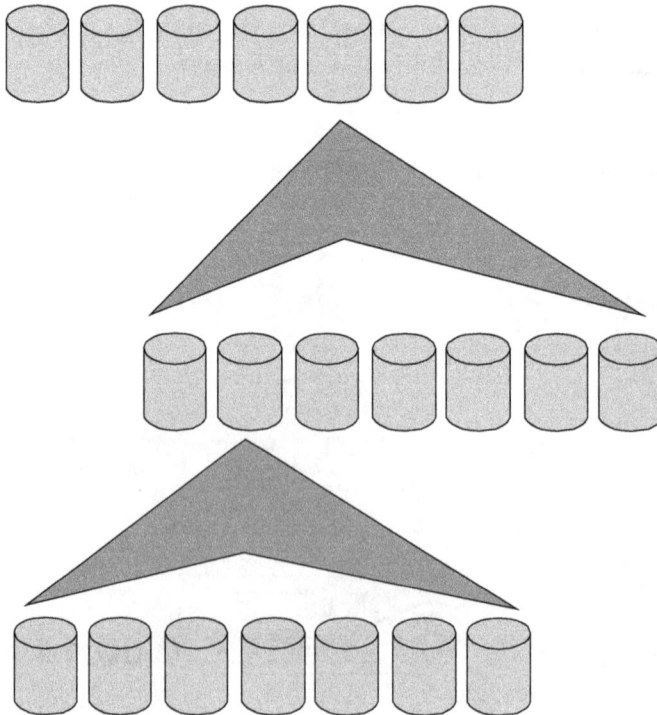

Figure 2.7.   Data Organized to Create Chunks.

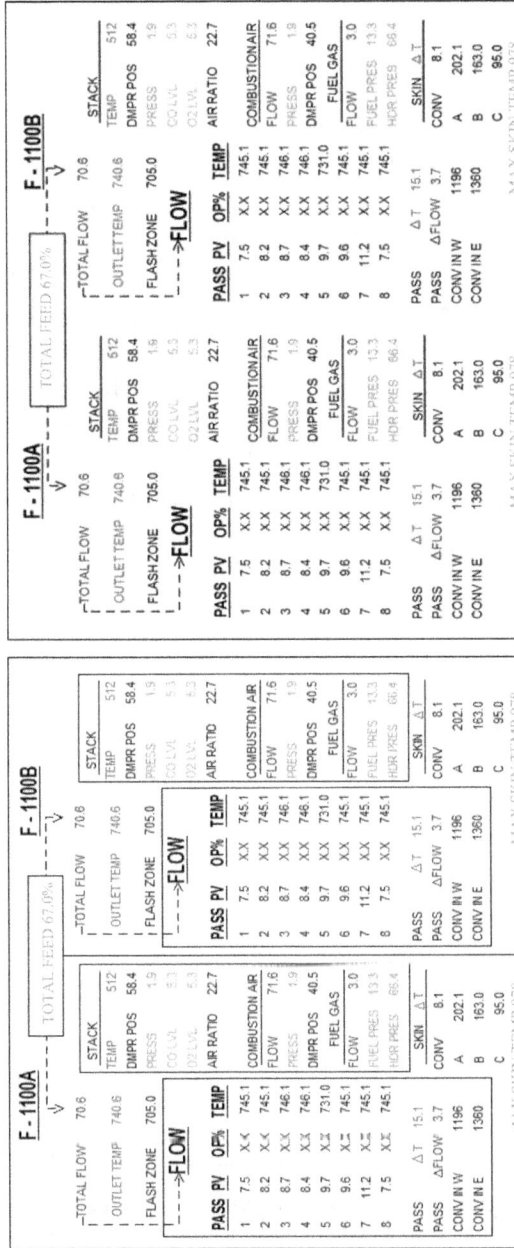

Figure 2.8. Use of Demarcation to Create Chunks.

Then within each part, there are chunks related to pass flows and combustion parameters. Processing and scanning are made easier because of the chunks that have been created.

2.4   MENTAL MODELS

Decision-making has been the subject of intensive research and has evolved since the beginning of my career, particularly on how people make decisions in real-world settings. One of the pioneers in this area is Gary Klein, who researched what he called naturalistic decision-making. The term was a contrast with classical decision theory researched in academic settings, wherein problem solving was a matter of defining the problem, developing alternatives, and selecting the alternative with the greatest benefit at the lowest cost. Klein's research into how people made decisions in the real world under stress revealed that alternatives were rarely generated or evaluated. So if they weren't choosing among alternatives, what were they doing?

Expert decision makers were engaging in analogical reasoning, using comparison cases to develop a course of action, similar to that in Figure 2.9 [6]. Events have a characteristic pattern or signature. Experts match that pattern to what they have seen in the past, and depending upon the success of past actions, either repeat them or develop a new set of actions. Once the situation had been assessed, the course of action was straightforward. Experts had better libraries of patterns upon which to compare the event at hand, in particular the ability to develop expectancies as to how the event should be playing out. Klein was of the view that it's just not that experts know more, but they see things differently.

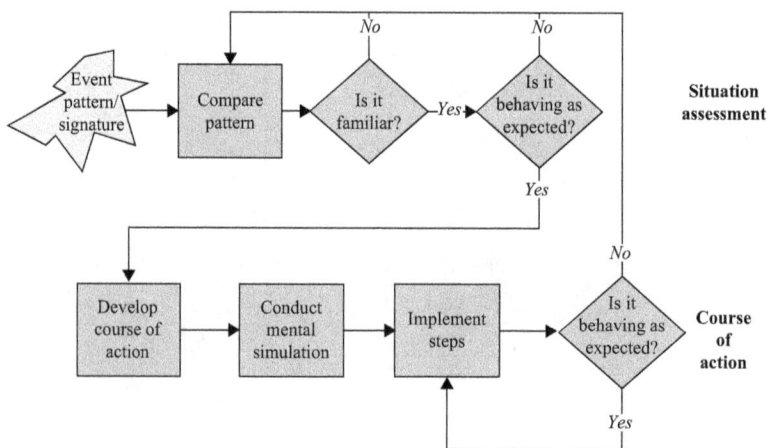

Figure 2.9.   Using Comparison Cases to Develop a Course of Action.

Improper situation assessment is often the point of breakdown in responding to an off-normal situation. Consider the operators at one plant with an automated safety shutdown system on their fired heaters. A separate light box had been installed for the shutdown system, so the alarms would not be lost on a scrolling alarm summary screen. The tiles on the annunciator were color coded in one of three colors: white for a fault in the system, red for a pre-trip condition, and green for showing that a trip had occurred. During a particular upset, the light box lit up as heater performance was swinging wildly. The heaters tripped, and the operator looking at the light box was fixated on the red pre-trip alarms, ignoring the green tripped alarms. Feeling he was getting the situation under control, the console operator asked the head operator for permission to have the field operator bypass the fuel gas chop valve. Permission was granted with concurrence from the shift supervisor. The field operator bypassed the chop valve, resulting in raw fuel gas entering the hot heater with no flame. The field operator was not on the heater deck for long, as he was blown off (without serious injury) when the fuel gas found a hot spot on the heater wall.

The console operator, head operator, and supervisor, all the three considered some of the best on the unit, would never put fuel gas into a hot heater with no flame. But all three agreed to do just that. How could three good operators make such a "mistake"? They failed to properly assess the situation. What they thought was occurring was not what was actually occurring. Their course of action was appropriate for the event they thought was happening, but disastrous for the real one. In this case, the shutdown system was new, and they likely had not yet developed good pattern recognition for a system with a somewhat questionable color coding scheme. However, as we've previously discussed, failure to properly assess the situation is a common thread in cases of "operator error."

A failed technology of the 1980s is worthy of mention. Expert systems were being developed that were intended to improve operator decision making. They failed miserably and at a substantial cost. This was largely a product of computer scientists who realized they could crunch a lot of data and provide the users with possible alternatives to the problem they were facing. However, if experts rarely considered alternatives, relying instead on analogies, providing more alternatives wasn't going to help. It was just confusing.

2.5 AUTOMATION AND RESPONSE

Another conclusion from Gary Klein's research is that the actions to be taken to compensate for a process disturbance are usually straightforward once the situation has been assessed. The problem in compensation is the number of actions that must be taken, often in a very short period of time. Automation can come into play in such a situation, having the DCS take a series of actions that otherwise would be taken by the operator. This creates the decision complexity advantage (DCA).

The theory behind DCA is that fewer complex decisions are superior to many simple decisions [7]. While a complex decision has a greater "refractory time" (i.e., the person will think about it longer), the net effect is that the total set of actions will occur more quickly. For example, consider a series of distillation towers in an environmentally conscious part of the country. Upon a major upset, the console operator needs to remove heat from the towers to minimize flaring. This is done by going to a series of five or six displays and putting the reboiler steam control valve in manual with an output of zero. Automation could enable that to be a single action, taking the heat out of the gas plant with one key stroke. Obviously, the operator is going to think longer before taking this action than one on an individual reboiler, but the net result is that the heat will be taken out of the gas plant more quickly in the automated situation.

One caution in automating such a system is the need for the operators to know what to expect when they activate it. A plant implemented such a system, whereby all the actions for a loss of compressor could be activated by the console operator. Fortunately for me (not the plant), a compressor trip occurred during the night of one of my visits. I asked the console operator if he had used the automation. "No," was his reply and he added, "I have never seen it work and wasn't really sure what it would do. So I just did it myself." Automation can greatly magnify human abilities, but trust in automation is a major source of research in the human factors community. (As an aside, this particular console operator made reference to having been the plant manager. I didn't get the joke but later found that it wasn't a joke and he really had been a plant manager. I then wondered, "How bad do you have to screw up to get demoted from plant manager to board operator." It turns out that he retired, started his own business, and wanted a steady income stream, so he became an operator again.)

2.6   CONCLUSION

We humans exhibit distinct characteristics in how we process information. Most operator errors are a result of our failure to either detect or properly identify the cause of process disturbances. If you want something to be detected, make the event stand out. Enabling problem identification requires the data to be structured into information with patterns that can be exploited for analogical reasoning.

3

Performance Shaping Factors

In the 1980s, we were typically asked to perform a human factors analysis of an event only when there was "operator error" involving the best operator on the unit. An analysis was done where, in response to a heater flange fire at one plant, the best operator in the division inadvertently shutdown the charge pump on the unit at the adjacent console. The analysis highlighted that cases of "operator error" are rarely one-dimensional or random; they involve the interplay of numerous variables that result in design-induced operator error.

The scenario was simple; several units had recently been upgraded to distributed control and moved to a single control room. A flange fire on a fired heater in Unit A was interpreted as a tube rupture, necessitating a unit shutdown. The uber-operator of the division was manning the board for Unit A. The operator went to the safety shutdown panel located on the wall to trip the charge pump in his unit. Instead of tripping the pump on Unit A, he tripped the charge pump on Unit B. The Unit B operator, who had moved over to assist with the Unit A shutdown, looked at his console to see a page full of red alarms. On discerning that it was a charge pump trip, he asked, "Is there anything in your shutting down that would cause my charge pump to trip?" After being told no, he restarted the charge pump on his unit.

How did this happen? In the move to the central control room, the charge pump shutdown switch for Unit A was not relocated, but this was not sufficiently communicated to the operators. Hence, in the operator's mental model, a shutdown switch existed. The shutdown switches on the wall of the

(*Continued*)

control room were an array of equally spaced switches for all the units with simple individual labels (no hierarchical labeling or demarcation). The operator scanned the shutdown switches in the area of the panel for his unit until he found one that was labeled "Charge Pump Shutdown." The stress of the situation caused him to overlook the "Unit B" on the label. He didn't notice that the charge pump failed to trip because, at about the same time, the shift foreman was walking down the pump alley shutting down all the Unit A pumps.

In this case, "operator error" was actually the interplay of training (not instructed that the switch had not been moved), interface design (poorly laid-out and demarcated shutdown panel), upset procedures/communication (who shuts down a key pump and doesn't tell the other people when it's done?), and automation (or lack of not having a charge pump shutdown switch).

The advantage as well as the disadvantage of human factors is the multi-variable nature of human performance. How well a person performs a task cannot be attributed to a single factor as it is the product of several variables and their interaction. That human task performance is influenced by several factors offers immense benefits to the human factors professional, as it means the professional can use a variety of tools to improve operator and system performance where needed. In many cases, deficiencies in one dimension can be offset or compensated by changes in another dimension.

This multidimensional aspect of performance has at least two downsides. First, too often deficiencies are not directly addressed as they should be. Additional training, adding alarms, or longer procedures become the quick solution to solving an operator performance issue, rather than thoroughly investigating true root cause and addressing the issue. Second, human performance issues rarely have simple and absolute answers. For example, for the question, "How many operators are needed to control this process?" the true answer is, "It depends—on the automation, the interface, the training program, and so on." Unfortunately, people want a definite answer, particularly when it comes to staffing levels and the associated payroll costs. In the 1980s, this led to a metric of control loops per operator. It was generally prescribed that an operator could handle 200 control loops. However, industry personnel routinely manipulated the data on their unit to achieve this target; they would often alter what constitutes a loop and add weightage to some loops more than others to ensure that the equation resulted in the desired staffing level. When I posed the idea of metrics to human factors professionals at Wright-Patterson Air Force Base, they were dumbfounded. They rightfully could not understand how staffing could be reduced to a single aspect of the system to be controlled.

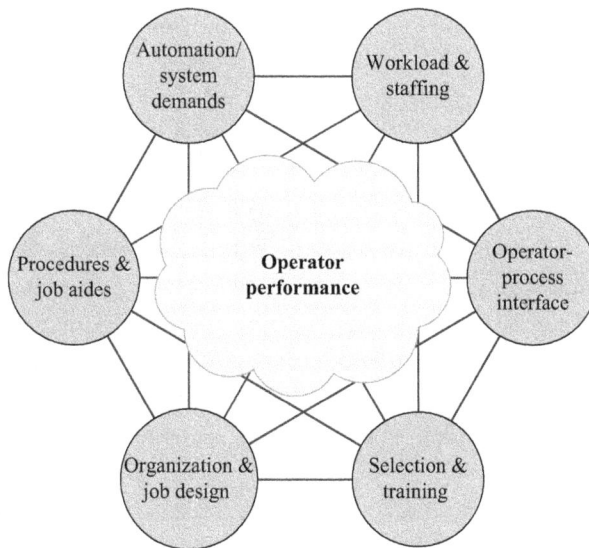

Figure 3.1. Performance Shaping Factors.

While there are a variety of ways to organize the dimensions of human performance, the six areas identified in Figure 3.1 are fairly common. The dimensions not only affect performance but also one another. Consider the interaction between interface and training. Display design involves a presumption of the knowledge of the user. The greater the level of information a user possesses, the less that information needs to be repeated on a display. Think about the design of an automated teller machine (ATM) interface. The only thing I can assume about the user is that they have a bank account, so the interface needs to be designed with an assumption of minimal knowledge of the user. Contrast this with a fighter pilot who has undergone maybe years of training before being able to fly a high-performance aircraft. It can be assumed that they have a fair knowledge of the system and that some of that information need not be displayed.

The design and analysis of any operator-process system consists of six dimensions, or performance shaping factors (PSF). If there is a deficiency in performance, then the cure lies in one or more of the PSFs.

1. *Automation/System Demands:* The answer to the original problem of what should the person do and what should the system do resulted in the creation of human factors and is embodied in what are known as Fitts' laws. I have included the demands imposed by the system on the operator, including how it varies with automation and alarming.

2. *Workload and Staffing*: This includes the number of individuals who are involved in the task(s) and their relative workload. Two items should be kept in mind: (i) both extremely high and extremely low workload can degrade performance and (ii) while increased staffing reduces individual workload, it increases team coordination demands.

3. *Operator–Process Interface Design*: This dimension consists of the organization, content, and formatting of the displays the operator uses. The biggest mistake made in designing solution to this dimension is taking into account only the format of the displays, and not realizing there are many more issues involved. Overall, this dimension embodies the right information, in the right format, at the right time.

4. *Selection and Training*: This dimension focuses on the individuals who have been chosen and how well they have been trained. The U.S. Nuclear Navy under Admiral Hiram Rickover emphasized this aspect of human factors probably more than the others. Reportedly, he wanted individuals to be able to adapt to whatever was thrown at them (that is, they should possess the ability to tackle any issue that comes their way).

5. *Organization and Job Design*: Any task is part of the overall job that has been designed by and functions within a larger organization. The organization establishes expectations and tasks to be performed, and the manner in which the work is organized into a job impacts how well it can be performed.

6. *Procedures and Job Aides*: While this dimension is not significant in more traditional models, use of procedures has been growing in such importance in the process industries (and is so poorly done) that I think it deserves to be treated as a variable on its own.

There are other ways to organize or model the PSF. Most other models would likely have a dimension on environment (i.e., light, heat, sound). I have omitted environment because, while its impact on the task needs to be assessed, it is not typically a variable that causes an "operator error" in process plants. Environmental factors in process plants tend to have a longer-term effect on job satisfaction. I have also omitted workstation design, such as the layout of consoles and control rooms. Again, this has less to do with causing error than efficiency, and other books exist that can aid the interested reader in this area.

While each of the individual human performance variables will be discussed in individual chapters, it is important to realize that any analysis will likely need to encompass most or all of the PSF. This is particularly true with investigations of cases where performance was not adequate for the task or "operator errors." Most likely, the incident is a function of one or more of these variables, which are potentially interacting.

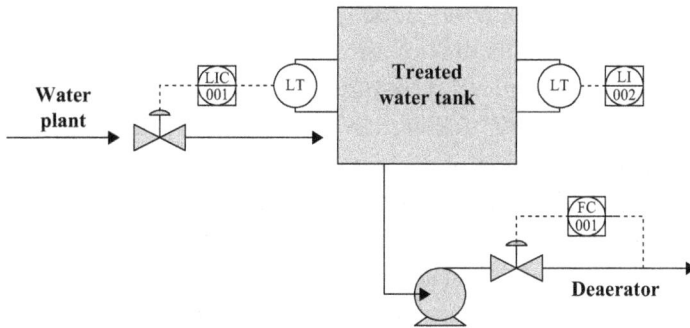

Figure 3.2. Treated Water System.

Let's look at an event that resulted in significant losses for a company caused by an "operator error." The event was simple enough: A level control valve failed on the treated water tank in the Utilities department of a major plant. A simplified diagram of the system is shown in Figure 3.2. When the level controller failed, treated water from the water plant ceased to come into the tank. The level continually decreased (for 8 h) until cavitation of the pump feeding the deaerator caused the spare pump to auto-start. Upon investigation of the auto-start, the loss of level was realized. The level controller was changed to manual mode and the valve was opened, but not before tripping all the boilers. The refinery and adjacent chemical plant were forced to shut down when the steam flow stopped.

Despite receiving a low level alarm on the redundant level transmitter, the operator said he did not recall notice an alarm indicating low level. The operator acknowledged that he was at the console, and the history shows a low-level alarm, but he maintained one did not go off that day. Was he lying?

An examination was made of the event history. It shows the level controller going above 90% and an alarm actuating on process variable high (PVHI) before going high-scale. The operator then lowered the set-point on the controller from 80% to 75%. A short time later, the redundant level indicator generated a process variable low (PVLO) alarm at 60%. At that point, the operator lowered the set-point further on the controller from 75% to 70%. Why would the operator lower the set-point for a low-level alarm? The only obvious answer is that he didn't think he received a low-level alarm, but instead thought it was another high-level alarm. The action taken in response to both the high- and low-level alarms was the same. The operator's statement about not remembering a low-level alarm is likely true. He doesn't remember because he thought there were two high-level alarms.

Let's look at this event from the model of PSF. This event clearly shows that influence of various factors led the operator failing to take the correct action.

- *Interface*: The alarm information was presented to the operator as a list, showing for each alarm its time of actuation, tag, descriptor, and alarm condition. The two alarms on level were distinguished in a long string of characters by two letters, with "PV" followed by "HI" or "LO." Although the operator had graphic representation of the system, only the controller changed color upon going into an alarm state (this was reportedly done to minimize confusion since the level indicator had a different span). The alarm summary that displayed the alarm had over a dozen standing alarms, so two more alarms did not stand out appreciably to alert the operator that something needs to be done.

- *Automation*: The system could have been designed to have a low-level select for the control valve. This might lead to overfilling the tank if the level failed low, but it would have only caused waste of treated water. Additionally, instead of two alarms, a single alarm with an alarm on deviation between the two would have better captured the potential for instrument malfunction than an alarm trigger for both levels.

- *Training*: What was the operator's only initial action to lower the alarm setpoint? Training should have emphasized the practice of confirming alarms with redundant indicators, at which point the two values drifting apart would have been identified.

- *Job design*: How does an operator go on for 8 h without knowing the status of one of the three parts in boiler operation (i.e., water, heat, and steam)? In the aerospace industry, there are examples of what is described as "forgetting to fly the plane," when the pilot neglects his or her fundamental task. In this instance, it seems the operator forgot to control the process.

- *Workload and staffing*: The operator had a very low span of control. Without much activity, it *might* be possible that the low level of workload resulted in lowered levels of operator performance.

What was the company's response to the incident? They added a low-low level alarm (PVLL) alarms to both instruments. As should be obvious, the problem was not that the operator failed to get an alarm but how he responded to the alarms he received. Additional alarms were not needed, but changes were needed in the operator–process system, in terms of training, displays, automation, or some combination of them. The cause and solution to the human performance problem is multivariable in nature. And the variables themselves interact with each other.

3.1   CONCLUSION

Operator performance is a function of multiple interacting variables. Identification of poor performance and improvement of operator performance require an examination of more than a single variable.

4

System Demands/Automation

I don't do much expert testimony, but I was involved in an arbitration case several years ago. A plant had built two new units and was asking their existing operators to absorb the additional work. The Union contended that this was unsafe. The Union presented their argument for half a day, which essentially came down to, "It's bigger, and there's more stuff now." As I pointed out in my testimony, there could be two valves a mile apart, but if there was considerable time between when they would have to be turned, then one person might be able to handle it. Conversely, two valves 10 feet apart that had to be operated simultaneously would require two operators. It is not the size of the "machine" that determines workload, but actions required by the operator.

The starting point for human factors is the demands that the system requires of the person, or in other words, what changes does the person need to make upon the system? There are two aspects of this human factors dimension that come into play. The first is global, the need to maintain a focus or perspective on the person's input to the system. The second aspect is more specific, that being how the design of the alarms and controls can impact what is required of the individual.

4.1 PERSPECTIVE

Many of the problems in human performance have a root cause of people taking the wrong approach to the design of man-machine systems. While the proximate problem may be too many alarms, excessive color, or confusing verbiage, the underlying cause is thinking that the design of systems for people begins with the system, not the people. It should begin with the person. This has been preached long before "user-centered" or "human-centered" became the flavor of the decade like "user-friendly" was before it. A system that considers human factors needs to consider the human first.

We may simply be the victim of how we in the West read. Consider the model in Figure 4.1. Information on a process is picked up through myriad sensors, which is displayed to an operator via some set of displays. The operator, in turn, processes that information to (i) do nothing, (ii) alter the display system, or (iii) effect a change on the process. The latter is done through either the control system or field personnel. Perhaps it is because we in the West read left-to-right that results in most displays being built by first considering the process, the hardware. Maybe cultures that read right-to-left are better at human factors, starting first with the operator. Or, maybe their diagrams would have the process on the right and they would be as bad as we are.

The proper approach to the analysis and design of man-machine systems is to start with the operator.[1] Specifically, what do they need to do? What output from them is required? A person exists in the system for some reason, generally to take an action. That action(s) should be the focus of the design effort. Given that an action is needed, what in the system will prompt the action to occur? For our situations, this will likely be something that they see on a display. What needs to be on the display to prompt the correct action? Now knowing what needs to be displayed, what process variables need to be measured to populate the display? It might be entirely possible that when this process is completed, there will be some variables left over. Why? Likely variables were added that are necessary, but not for the operator. These may be measurements for rotating equipment or process engineering personnel. While they may need to be on a display somewhere, obviously they should be given the emphasis that those variables critical to operator action demand.

When I have challenged display designers about including data that has no effect on operator action, I am given one or both of the following arguments: (i) it might be nice for them to know, and/or (ii) it can't hurt. Knowledge is limitless, so nice to know could be anything. It provides no discrimination on whether something should or should not be included. And it can hurt. Not only does unnecessary information increase the visual noise of the displays, when repeated it can result in a bloated display system. This leads to unnecessary navigation, complexity, and makes revisions more difficult.

---

[1] I still use "man-machine," which is what I learned in college, just as I still say mankind. The use of human in human-machine systems is too politically correct for me in that the root word of human is, of course, man. I use operator-process interface if gender neutrality is desired.

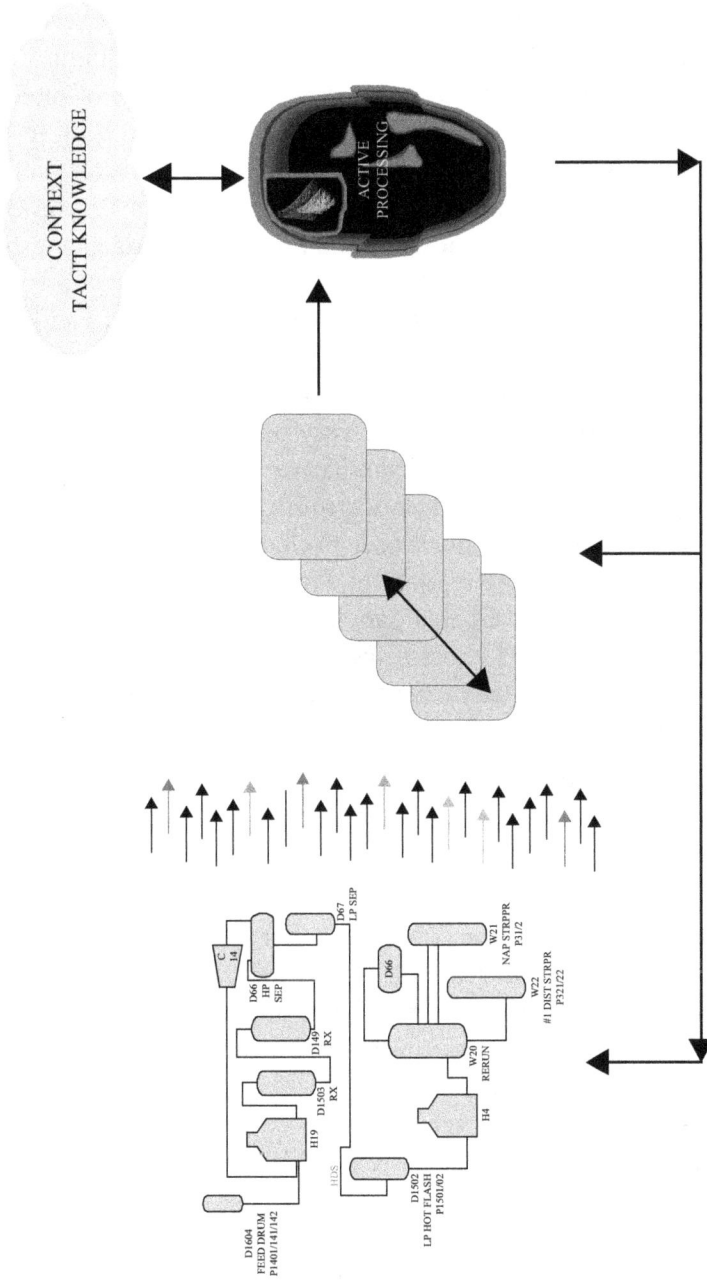

Figure 4.1. Model of Operator Process Interaction.

Even when the person is nominally considered by system designers, too often the designers lack either the necessary analytic tools or focus to create a user-friendly design. The lack of analytic tools manifests itself in a folk psychology or overreliance on user preference. Folk psychology is treating yourself and your views as universal to all people, "If I like and can use it, everyone must like and be able to use it." I run into folk psychology frequently in regards to how many alarms an operator can manage. People think of one alarm per minute as unmanageable. Certainly if that went on for too long it would be. But stop reading now and count to 60. Most likely, that either seemed like a long time or you got bored and stopped. Humans are not very good at time estimation and perception. The novice thinks 1 min sounds like a short time. While it is for some things, it is not always.

Over-reliance on user input has subtler flaws. In an effort to address human factors issues, designers seek out the end users of their creation. This is generally good. The problem arises when the extent of the input is asking what the user would like. We humans are actually very poor at defining what we need. This is often the result of a bias toward what we have or have seen. The result can often be an unexpected missed opportunity.

We encountered the problem of user preference in a plant converting to distributed control. The control panel in Figure 4.2 was mounted in a control room. The operators loved this panel. They told the display designer to duplicate it on the distributed control systems (DCS). And had he done it, he would have been giving the user exactly what they asked for. However, a few questions proved illuminating.

Q: What is this panel used for?
A: We use it to add or adjust one of four fuel types to one of the quadrants in the boiler.
Q: How is that done?
A: If we are adding a fuel, the control for that is on the left side of the panel in the associated corner of the boiler. We open that valve. Then we adjust the flow controller (not on this panel, but a few feet away) while watching the pressure (also several feet away) to ensure that it doesn't go too high or low, resulting in a trip.
Q: So to use the panel, you need the flow controllers and pressure indicators that are not currently on the panel?
A: Exactly.
Q: What is the right half of the display used for?
A: That is used if a fault is indicated by one of the lights on the right side lighting up.
Q: So you don't need to look at the right side unless one of the lights is lit?
A: Right.

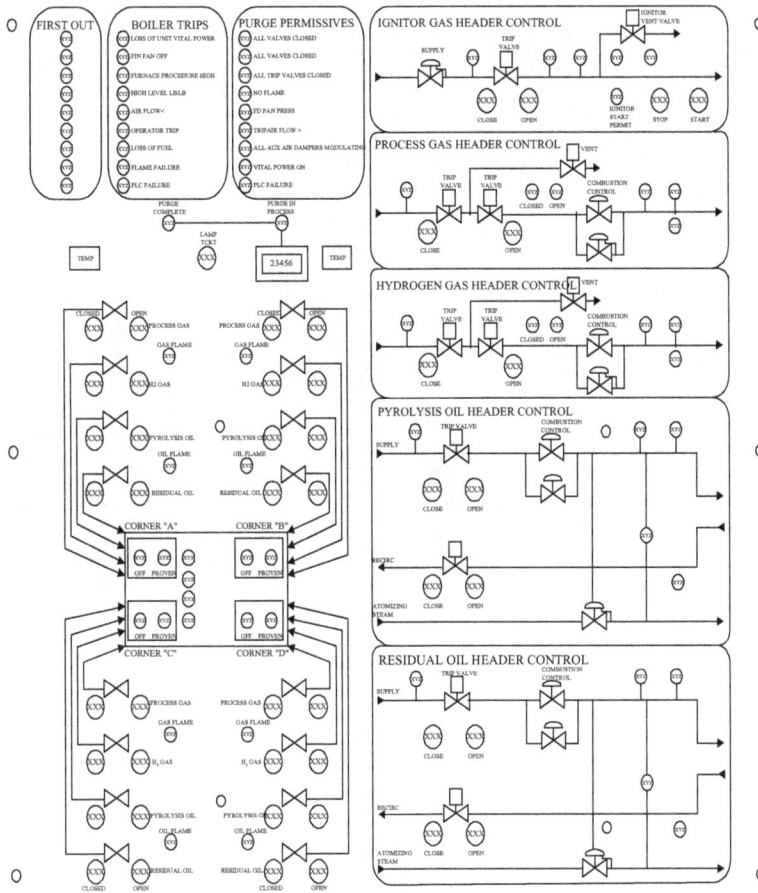

Figure 4.2.   Boiler Firing Panel.

Hopefully at this point you will realize that duplicating that panel on the DCS would have been less than optimal. The operators would likely have thought the display was good, because it looked like what they had. What it wouldn't have had was the controls and other information the operator needed to use it. It didn't have everything needed to perform the task and contained information that was only infrequently used and only when a cue was available. Just getting user input is insufficient for good user-friendly designs.

Well-intentioned designers often fall short because they think their job is to present information. There is a big difference between just present-ing information and optimizing information transfer. This lack of empha-sis in maximizing information transfer can be seen in displays, procedures, and training manuals. We encounter it in our everyday lives with rambling letters/memos ("get to the point"). Visual information transfer can be

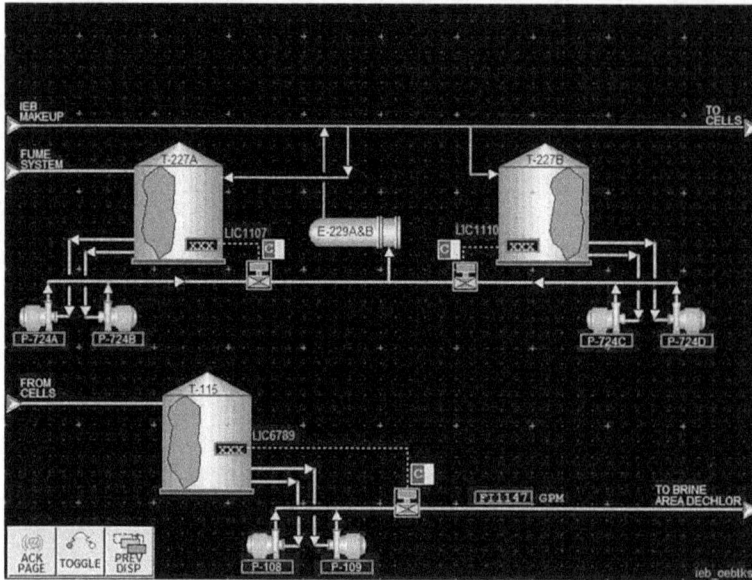

Figure 4.3.   Not an Atypical Display.

thought of as bits/in^2. In display design, the actionable information divided by the display area gives the information transfer. In text, the information divided by the number of characters defines the information transfer. Sparse displays and wordy manuals have low information transfer. Consider the display in Figure 4.3. It has the status of six pumps, three levels, and one flow for the entire display. Well over 50% of the display is not utilized. The information transfer is very low.

Consideration of human factors includes the evaluation of visual information in terms of bits/in^2. In examination of a display, a keyboard, a paragraph, what is the information transfer rate? Why do we encode information, such as through the use of color? It increases the information transfer—more bits per square inch. Unfortunately, maximizing information transfer requires more effort than just throwing symbols on a page. It is far easier to simply use multiple displays rather than work to put the same data on one. It is far easier to write training manuals as a stream of conscious thought, rather than edit and re-write to maximize the value of what is said.

Human factors is not just approaching the problem with the correct paradigm, but also understanding that we have control over some of the demands the system will place on the user. In process plants, this largely falls into two areas: control changes and alarms. The control system design will impact how and when an operator must take action and the alarm system creating requests for the operator to take an action.

4.2 CONTROL SYSTEM IMPACT

I am not a control engineer, although I often play one during alarm ratio-nalizations. However, anyone who observes console operators quickly under-stands the impact on them of the design of the control system (or lack thereof). Both basic process control and advanced process control have implications for operator performance. I was with an operator on a delayed coking unit in Southern Texas who had to make repeated control changes throughout the afternoon. Near dusk he mentioned that he was turning on the advanced control package. I asked the obvious question, "Why hasn't it been on?" I received a slightly mocking reply of, "It's hot outside, and it doesn't work when it's hot." It seems that the units rates had increased since the advance process control (APC) was installed and could no longer take the heat of the day, leaving the operator to make the changes.

At one time, the person was the control system. If a level got high, they opened a valve or started a pump. Temperature too low, add some fuel to the furnace. While there are still instances of this, by and large humans do not control the process, the control system does. Humans monitor the control system. Dr. Tom Sheridan, a man I greatly admire, coined the term "supervi-sory process control" to distinguish the role that the person was now playing in the system.

This change has altered not only when the person interacts with the system, but the human factors associated with process control. The skills required of the operator and the information that is needed to be presented are not the same as automatic control has advanced. Such is the interactive nature of human factors. The ground breaking work in this area was by Jens Rasmussen in the 1980s that demonstrated the hierarchical nature of human behavior [8]. Rasmussen's model is one of three levels of behavior, triggered by signals, signs, or symbols, as in Figure 4.4. At the lowest level, an input is associated with a single output (e.g., a level is high, so open a valve). This level of human behavior has largely been replaced by single loop control, such as PID controllers. In the next level in the hierarchy, a set of inputs results in the initiation of a series of actions, trig-gered by some sign. This level of behavior is akin to and being replaced with advanced process control. The highest level of the hierarchy requires the use of symbols and activation of mental processing on goal states to develop a course of action.

We had the opportunity to observe firsthand the impact on the operator as their primary role/function moved up the behavior hierarchy. A batch plant had been increasing their automation to the point where, if everything went as designed, the operator had three tasks: (i) select the batch, (ii) initiate the batch, and (iii) transfer the batch upon completion. However, things did not go as designed all the time. When this happened, the batch program would go

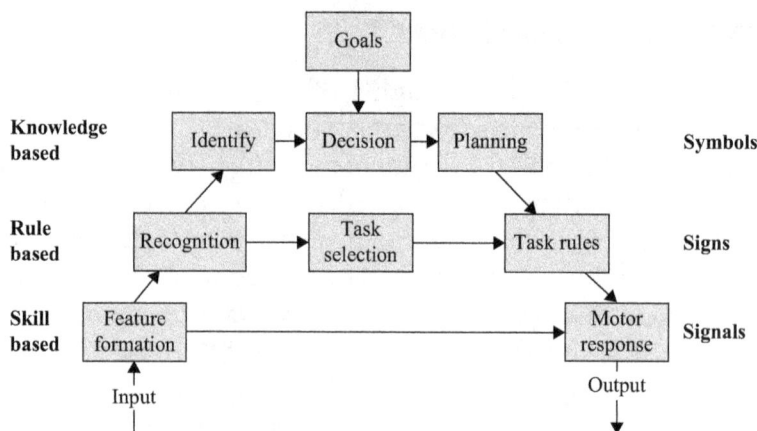

Figure 4.4.   Hierarchy of Behavior.

on "hold." The operators would get a message whose intent was to empower the operators to remove or resolve the source of the hold.

Unfortunately, the operators were lost. If an engineer were available, they would be called into the control room to assist. If they weren't available, they would be called at home, at any hour. Well, this was intolerable! In discussing why the operators couldn't trouble shoot the hold, the problem crystallized rather quickly. The displays showed the operator the process and process variables. But the operators weren't controlling the process, the batch program was. And that was invisible to them. Essentially, they had become supervisors of the batch program. Automation had taken over the skill and rule based behaviors. The operators were needed for knowledge based behavior. They needed to see what the batch program was doing and what it expected to be done, its goals. This was accomplished by adding the batch step information in Figure 4.5. The operators could then understand the range of options that might not have been satisfied, creating the hold.

Automation not only changes the information and skill requirements, it can significantly degrade in the operator the very skill that it is performing. As the advanced control does what the operator once did, the lack of practice on the part of the operator results in their skills being degraded. While this has not been quantified yet in the process industries, many plants have recognized the problem. At one facility, they would turn off the advanced process control for weeks at a time to enable the operators to practice. The cost of not having the advanced control operational was in the tens of thousands of dollars per year. And they still couldn't be confident that the operators had retained the crucial skills. They only hoped it had done so.

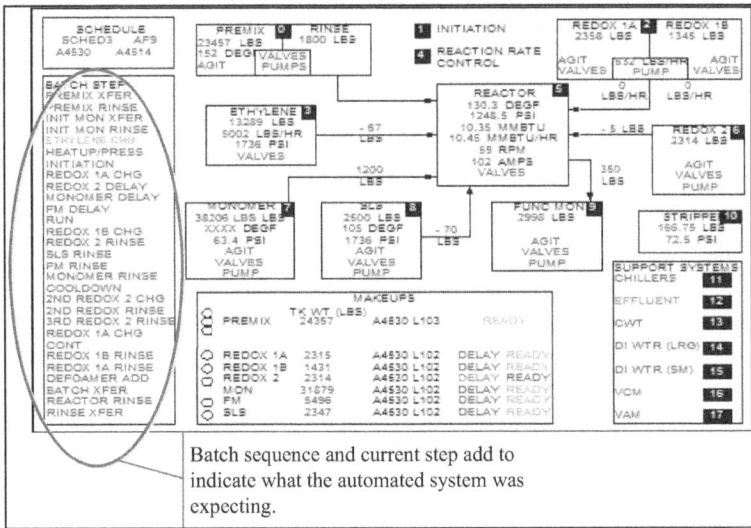

Figure 4.5. Display for Higher Level Control.

4.3 ALARMS

I am not going to go into great detail on alarm management. There are a variety of books on the topic, parts of which I think are wrong, or at least they make it far more complicated than it need be. However, alarms are a major generator of workload and system noise for process operators. Alarms are requests for the operator to intervene in the system, so every alarm should be a demand of the system for the operator to act. How many alarms can an operator act upon in a short period of time?

One of the first set of targets for alarms was published in 1999 by the Engineering Equipment and Materials Users Association. To their credit, they state that the targets are based upon industry experience and not theoretically based. Other documents since then (ISA18.2, Alarm Management for the Process Industries) have largely kept the same targets. The targets include [9]:

- Average alarm rate < 1 per 10 min.
- Alarms in 10 min after upset < 10.
- Average standing alarms < 10.
- Average shelved alarms < 30.

The Center for Operator Performance was curious about this issue as well. Were the published targets good limits? Dr. Craig Harvey of Louisiana State University conducted a series of experiments with students and then process operators. Exposing the student subjects to five different conditions of 1, 2, 5, 10, and 20 alarms in a 10 min period, he found that a queue did not develop until 20 alarms in 10 min (Figure 4.6). When repeated with pipeline controllers and process operators, the effect was not seen until 30 alarms in 10 min [10]. Clearly the published "limits" were conservative at best.

The role and effect of prioritization on performance can be seen in the study results. In the highest rate condition, the subjects sacrificed response on the lowest alarm priority rates to handle the higher. This is what you hope happens and why alarms are prioritized.

Prioritization aids the operator in the determination of what has to be done now and what can wait. While this is generally accepted, we have seen an amazing assortment of distributions for alarm priorities. Alarm "experts" will cite the 5-15-80 distribution as the "right" one. Is it? Consider the following thought experiment, what would be a variation of a paired comparison approach. Operators could go through all the alarms and be asked to separate them into two equal piles of important alarms and less important alarms. The important pile could be subjected to the same sorting of important and less important. This can be continued for each pile considered the "more important." At the end, various piles of alarms would remain with increasing levels of "importance." Think about a situation where an operator had 100 possible alarms. This is convenient as the sorting readily converts into percentages. The first split is into two piles of 50. The "important" pile of 50 is further split into two of 25, and so on, just as in Figure 4.7. The result would be the order in which the operator should attend to an alarm.

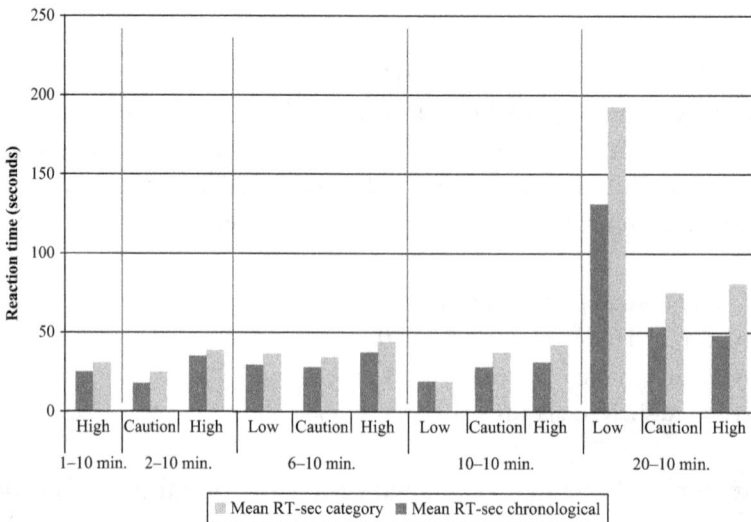

Figure 4.6.   Reaction Time for Different Alarm Rates and Displays.

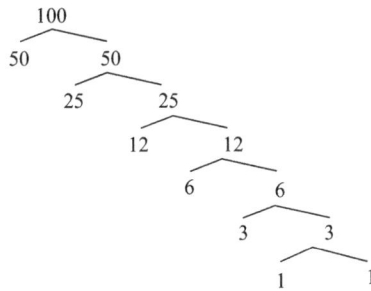

Figure 4.7. Paired Comparisons and Alarm Priorities.

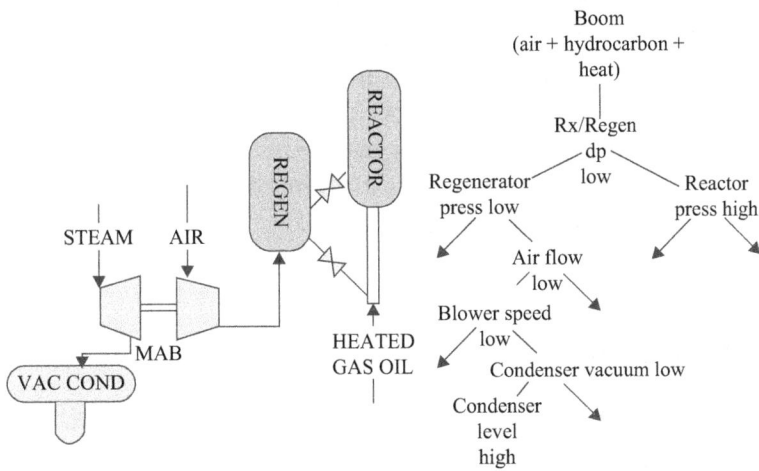

Figure 4.8. Alarms Sequence on Fluid Catalytic Cracker (FCC) Reactor/Regenerator.

Obviously, this is not practical if there are hundreds of alarms, but the basic concept still holds. The number of alarm levels would determine the percentage of alarms expected at each level. We find that for a three-level system, 50%/35%/10% works as a good configuration distribution. However, you should be able to see that other combinations would work if reducing to three sets, say percentages of 75%/20%/5% or 78%/12%/10%. So the "right" answer of 5-15-80 is reasonable, but not inviolable.

Selection of what to alarm is not that hard, it results in an alarm that prompts a unique operator action, with priorities escalating as alarms are not attended to. The escalation effect is often overlooked, resulting in a skewing of priorities to the high side. In Figure 4.8 for example, air is blown into a vessel where it comes into contact with a catalyst to burn off excess carbon. The cleaned catalyst mixes with gas oil before going into a reactor vessel at around

1000°F. Air and hot hydrocarbon could potentially mix if pressure were lost in the regenerating vessel. Initially, the operators had the high surface condenser level as a critical alarm, since it could ultimately lead to an explosion. However, before that would occur, there are five potential intermediate conditions that could potentially alarm, and so the initiating condition need not be the highest priority. Subsequent alarms need to be a higher priority to reflect that the initiating event was not resolved. It should also be clear that all the conditions need not be alarmed. They could if each had a unique action associated with it.

Ironically, adding additional alarms has only marginal value in human performance. The greatest performance benefit occurs between having no alarms and having an alarm. Increasing specificity from the alarm system does improve performance, but not to the degree that having the first alarm does [11]. Just letting the operator know that something is wrong and needs attention has the greatest benefit in operator/system performance. So at what point does adding alarms reduce performance by increasing workload? That answer has not yet been determined, but there is a point of diminishing returns, and it certainly would be worth knowing.

4.4   CONCLUSION

The starting point for operator performance is understanding the demands required of the operator by the system. However, some of those demands can be altered through system design and automation.

5

Workload and Staffing

A large portion of our work is in workload and staffing evaluations. We often utilize a technique called job sampling, which involves our observation of an operator performing his job in 4 h increments of time. A criticism is that the individual knows he is being watched and will either be at his best or potentially "make up work" to appear busy. While the latter has certainly happened, and we count on the former to ensure our results are conservative, it is not universal that people act busy when they are watched. There have been repeated instances in which the individual being observed has fallen asleep.

A new employee of ours asked me for guidance on what he should have done when the operator he was observing dozed off, he told me, "I didn't think he should be sleeping, but I know we try not to do anything that would influence the results." I suggested making some noise. It generally works for me. It is less effective if the operator asleep is not the one you are observing. The operator under observation may then have to deal with the embarrassment of hearing their co-worker snoring.

We all have some natural appreciation for workload. Comments such as "That was a hard day," "That looks like a hard job," and "That job is a piece of cake" are often heard. However, our natural appreciation breaks down, as much folk psychology does, when we are confronted with creating a systematic analysis of human performance. "Every day is different," "this job is the hardest in the plant," and "there are days I don't get to eat lunch" are all common responses when an attempt is made to measure workload and extrapolate it to required levels of staffing. While I believe that there are days when every operator doesn't have time to eat lunch, the question is how often do those days occur? Once a week, a month, a year, or a decade?

It turns out that even experts aren't good at judging how much work an individual can handle. A case of how experts can underestimate what we humans are capable of doing was seen by the U.S. Navy. Current technology enables cruise missiles to be controlled after launch. These missiles are no longer fire-and-forget, but can be retargeted mid-flight based upon changing situation dynamics. The Navy's original estimate, an acknowledged "best guess," was that an operator could handle four missiles. It turns out they were wrong by about a factor of three. Little difference was seen in controlling either 8 or 12 missiles; however, significant degradation was seen when attempting to control 16 missiles [12]. Fortunately, the Navy decided to quantify the workload and performance for this task, thereby enabling better personnel utilization.

Defining what we mean by workload is necessary. Within the human factors community, workload is not a single aspect of performance but a term encompassing a variety of human performance measures. At its most basic level, workload is broken down into physical workload and mental workload. Within each of these, there are numerous aspects that are studied and quantified.

5.1   MENTAL WORKLOAD

Mental workload is largely of interest when the workload of console operators is considered due to the more cognitive nature of their job. Does a specific console operator have enough spare mental resources to process the information needed to assess, diagnose, and respond to the problem? In the short term, mental resources for any individual are fixed [13]. There is only so much mental resource we have at any given moment to apply to information processing. In the long term, mental resource capacity can be altered, which will be discussed later.

A typical method for assessing mental workload that highlights the fixed resource nature is a dual task test. In these studies, an individual is given two tasks to perform—the one of interest and a secondary task [14]. For example, an individual might be placed in a car simulator (the primary task) and asked to subtract backward from 1000 by sevens (the secondary task). As the primary task difficulty is increased (nighttime, rain, narrow roads), there comes a point when the secondary task begins to degrade (longer pause between responses, more errors). This happens because the primary task has used all the available mental resources, and insufficient resources exist for the secondary task, as illustrated in Figure 5.1.

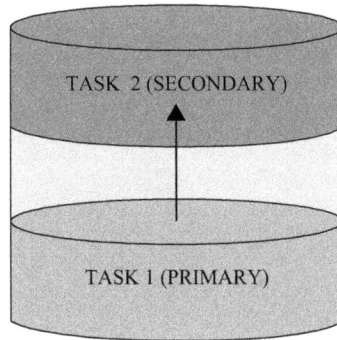

Figure 5.1. Mental Resource Usage.

Mental resources will change over time, by expanding and contracting. Training can obviously expand mental resources (this is an oversimplification of a complex concept, but it should suffice for this discussion) [15]. Failure to utilize mental resources can result in their contraction. This is why the console operator's workload needs to be kept at a reasonable level during steady-state operation, so that mental resource are preserved. In addition, it is possible to dedicate certain mental resources to performing specific tasks, which is accomplished through repeated practice. This creates a situation where that task can be performed with little impact on available mental workload reserves, a phenomenon known as automaticity [16].

A technique for assessing the mental workload required for various tasks is called subjective workload assessment technique (SWAT) [17], which was developed by the U.S. Air Force and adapted for process plant use. This analysis requires individuals to rate the effort of a task along three dimensions: (i) time, (ii) mental effort, and (iii) stress (Figure 5.2). From the ratings along these dimensions, the total workload can be calculated. It should be obvious how system design can impact these dimensions and, consequently, total workload.

- If rote tasks are automated within the distributed control systems (DCS), the operator has more time for other tasks, then and the workload is less.
- If the interface has been designed to minimize the time hunting and searching for the key variables, than workload is less.
- If the operator has trained on the upset so they are confident of their response, thereby lowering stress, then workload is less.
- If the formatting of the displays make it easier to solve the problem, then workload is less.

We were able to see the impact of confidence on stress firsthand, although formal training was not exactly the cause of confidence. We used SWAT to

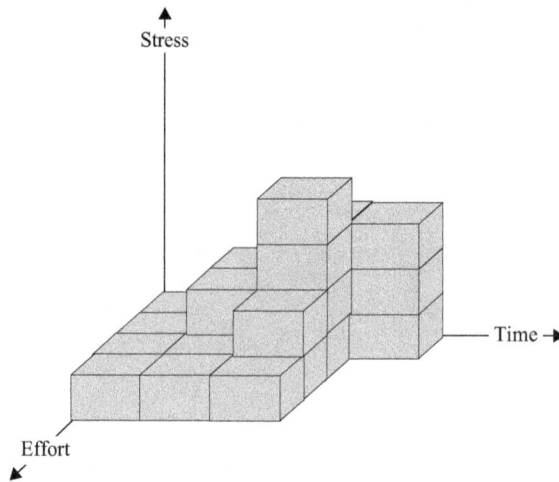

Figure 5.2.   Measured Dimensions of Subjective
Workload Assessment Technique (SWAT).

assess operator mental workload for a series of upsets at multiple units in a refinery. To everyone's surprise, the operators of the sweet crude unit (slow dynamics, very stable) had higher mental workload demands than the operators on the hydrocracker (a high pressure, high temperature) unit. The latter had poor reliability, barely able to stay running for two weeks in a row. The hydrocracker was what everyone thought should have the highest mental workload demands, not the slow dynamics and highly reliable sweet crude unit. However, upon further investigation, we found one underlying cause of the low hydrocracker workload to be the unreliability of the process, or more specifically, that the hydrocracker operators practiced upsets in huge numbers. To them, upsets were not challenging tasks. The sweet crude unit operators, on the other hand, thought any upset would be a catastrophe.

The Center for Operator Performance conducted a study that successfully utilized SWAT to measure the impact of display design. Two different graphic designs were utilized: a state-of-the-art graphic and an intentionally "bad" graphic (excessive color and detail). The subjects, using a simulated debutanizer, were tested using a variety of techniques. SWAT was successful in detecting performance differences between the two interfaces [18].

While excessively high workload is an area of concern, the negative impacts of low workload cannot be dismissed. It has long been known that low levels of workload can adversely impact human performance. A study that involved electric shocks to rats in the early part of the twentieth century found that both high and low levels of stress negatively impact workload, creating an inverse U relationship termed the Yerkes–Dodson effect (Figure 5.3). The

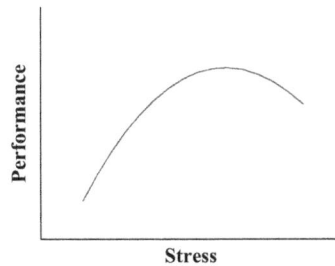

Figure 5.3. Yerkes–Dodson Effect.

results have been expanded to show the relationship with a variety of stressors other than electric shocks, including workload [19]. Keeping workload low to ensure that an individual can perform at their best can backfire if the workload is too low. This discussion is continued in the discussion of workload of console operators.

5.2 PHYSICAL WORKLOAD AND STAFFING

It is a common perception that physical workload is thought of in terms of exertion. While exertion can be calculated and is an issue in some industries, it is not of significant concern in process plants. In the process plant setting, physical workload corresponds to the observable behavior, the tasks or outputs of the operator subsystem. It is physical workload that will determine the staffing required, as it is people who will perform the tasks required by the process and organization. The issue in this context is to understand the human output of the system and what actions are required of the operator?

Trying to determine the staffing requirements of a process unit is made difficult by the different modes of plant operation and fundamentally different types of positions. Process plants have five basic modes of operation: steady state, upset, start-up, shutdown, and down (Figure 5.4). Start-ups are scheduled and are of relatively short duration; staffing needs can generally be handled with operators working overtime as necessary. However, the first three modes (steady state, upset, and shutdown) create problems in determining how many operators are needed to run a process unit. This is further complicated depending upon whether the operators are console operators or field operators. During an upset, console operators tend to be reactive, as opposed to during steady state, where they are presumed to be working proactively to prevent upsets. Field operators, conversely, tend to be reactive in steady state and proactive in an upset. By proactive and reactive in an upset, what is meant is whether the operators know what they need to do at the onset of the upset.

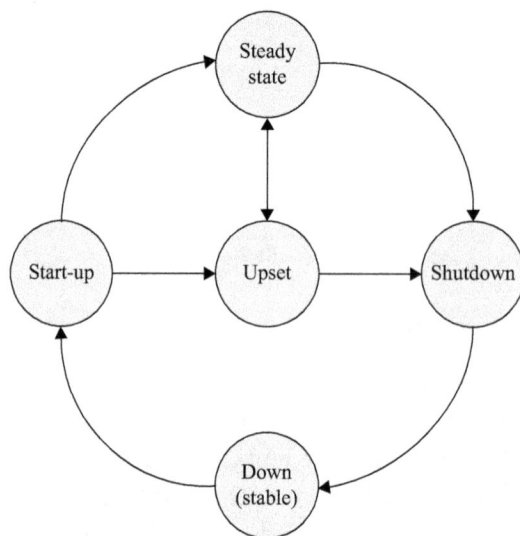

Figure 5.4.   Modes of Process Plant Operation.

Table 5.1.   Staffing determination by job type

Position	Mode of operation	
	Steady state	Upset
Console	Staff for	Design for
Field	Design for	Staff for

Console operators will need to *react* to how the process responds to the field operators' set of prescribed actions (is it cooling fasting enough, are levels stable), or do they need to intervene?

An observation of hundreds of operating positions clearly shows that the approach to staffing varies with the job type and plant mode, as given in Table 5.1. The determination of minimum staffing is different for field and console operators due to the different nature of the jobs. For the console, steady-state workload should be the determinant of staffing, whereas upset response demands will set the field operator minimum staffing.

5.2.1   Console Staffing

Console operators are recognized to have dramatic swings in workload between steady state and upset modes of operation. They are often referred to as "bored operators" with the low workload that occurs when a unit is running smoothly. Yet during an upset, it is not uncommon for the console operator to be expected to handle 50 or more alarms per minute, in addition to controlling/directing the response to the upset. One operator described himself as "I look like Ray Charles on cocaine" during an upset. It is somewhat surprising, then, that the staffing of the console should be based upon steady-state workload demands.

The reason consoles are staffed to reflect steady-state demands is that low workload can be as detrimental to operator performance as high workload. Recent research indicates that sustained periods of low workload inhibits the individual's ability to handle high workload periods, in essence, shrinking their mental workload reserves to match the "normal" low workload period [20]. As such, staffing should be set to ensure adequate workload for steady-state operation, thereby keeping the operator engaged and alert. Typically, staffing the job to be able to handle the upset results in the low steady-state workload associated with the "bored operator," which in turn actually limits their ability to respond to upsets when they occur.

If console staffing is based upon steady-state workload, how then is the console operator going to handle the sudden increase in workload when an upset occurs? Modern DCS have significant flexibility in how they are configured, enabling both a wide range of options for presenting information and the ability to automate rote sequences of tasks that the operator performs. In addition, while the upset will have significant response requirements, those requirements for handling upsets can be reasonably well determined through a variety of analytic techniques. The flexibility in DCS configuration should be utilized to ensure that the staffing required for steady state is sufficient to safely handle anticipated upsets. Designing the operator–process system (OPS) so that the steady-state staffing level can handle upsets involves manipulating one or more of several areas:

1. *Automation*: Any rote tasks that the operator must perform for a given upset should be performed by the DCS. For example, removal of heat from a series of towers should be a single command for the console operator, with the DCS carrying out the individual steps (e.g., putting reboiler temp controllers in manual with zero output).

2. *Alarm system*: At this crucial time, the ability of the alarm system to provide effective warnings, particularly of malfunctions independent of the initial upset, is paramount. This requires a good base alarm system, potentially combined with (a) logic to suppress those alarms whose actuation would be "normal" for an upset and/or (b) aggregation of alarm information into higher order (system level) alarms.

3. *Display system*: Rapid access to critical controls and information, presented in a format that enhances its use, is needed to speed operator processing and response in a time-critical situation. This will likely require critical controls and

information being condensed onto a fewer number of screens than are currently used.

4. *Training*: Since the basic aspect of most upsets are known, the operators should be trained for rapid identification and response to those upsets. Given a set of symptoms, the console operator should have an idea of what the cause might be and how to verify it. A key characteristic of expertise is the ability to conduct mental simulations of events [21].

5. *Support systems*: Decision support and situation awareness (big picture) need to be applied as needed. This may come from procedures or other individuals, but ensuring that all the operators are working from a common script in response to an upset is essential [22].

6. *Workspace*: The layout of the control room needs to be such that critical communication links are enhanced while distractions are minimized. Noise and traffic flow need to be managed by the design and arrangement of workstations and the control room.

Any workload measurement technique must account for the various tasks performed. The tasks can generally be thought of as belonging to one of two categories: job-dependent tasks and system-dependent tasks. Job-dependent tasks are typically set or done to meet management needs, such as filling out log sheets, setting work schedules, and periodically monitoring the system. These tasks are typically discretionary in nature when they are performed and, as such, are much easier to accommodate. Job-dependent tasks impose workload independent of the activity required to control the system; they are constants for the job and not the span of control. System-dependent tasks are those related to control of the process. They typically are somewhat random in nature, being initiated by the process, and must be attended to in a short period of time. Responding to alarms, making control changes, sending/receiving instructions from field personnel, and optimizing the process are the critical determinants of system-dependent workload, which increases with increasing span of control.

Since consoles are characterized by a large number of short duration tasks, console workload evaluation needs to reflect both the number of tasks to be performed and their duration. A common method to measure workload is in terms of percentage of utilization, as in, what portion of the shift was the individual engaged in job-related tasks. This is the sum of the duration of all the tasks for a given period. In conducting workload evaluations, we were surprised to discover that console operators average well below field operators in time on task (approximately 50% for console operators versus 75% for field operators). However, we noticed that our subjective descriptions did not match the objective data. A console at 60% utilization would often be perceived as busy, and a field at 65% utilization was described as "dead." In digging into why the objective and subjective did not match, we saw that the console workers had far more tasks per hour than their field counterparts.

The result was development of a way to combine the two metrics: time on task and number of tasks, into a single metric, mean time between tasks (MTBT). This is the available time (total time minus time on task) divided by the number of tasks. We refer to this as the degree of busyness. Consider two positions: one spent 45 min out of an hour doing four tasks versus another that spent 30 min on 10 tasks. Which one had higher workload? It depends on how the tasks are defined. However, using mean time between tasks, the first position had a MTBT of 3.75, while the second was 3.0. For MTBT, a smaller number means a busier job, with less time available to transition from one task to another. In this example, the second position was busier. Use of this metric allows an equivalency to be determined. For example, a position spending 40 min out of an hour on 10 tasks would have the same MTBT as a position that spent 30 min on 15 tasks. For our console and field operators, the difference in workload as reflected by MTBT shows the average console operator to be busier (MTBT of 1.1) than the average field operator (MTBT of 1.7), despite the field operators spending more time on job-related tasks.

An example of the MTBT metric, that of pipeline console positions, is shown in Figure 5.5. The time on task and number of tasks for each sample of our pipeline controller sample averages is plotted. The diagonal lines are equivalency curves reflecting equal MTBT on the line. The average MTBT is approximately 1 min. If an operator has a MTBT of greater than 2 min, we would describe the position as underloaded. Those positions that have less than one-half a minute MTBT are those that we would describe as overloaded.

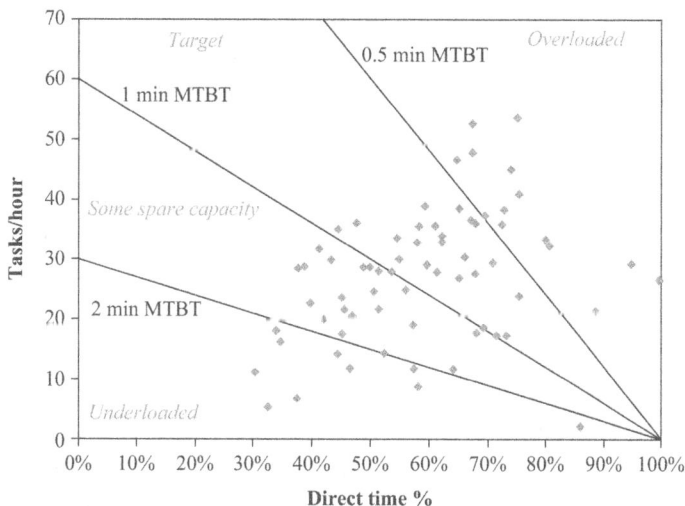

Figure 5.5.   Mean Time Between Tasks (MTBT) Example.

However, the one-half minute mark is not a fixed limit, but should be treated like the red-line on an automobile's revolutions per minute (RPM) meter. You can exceed it; you just don't want to spend a long time there. If the MTBT is too low with little time to rest and recover between tasks (i.e., the operator is too busy), either reduce the number of tasks or reduce time on task. Alarm and control system improvements are likely to have a greater effect on number of tasks, whereas improvements in administrative tasks or display design will likely have a greater impact on time on task.

The steady-state staffing needs to be able to manage an upset when it occurs (and it will occur). This is where the design of the OPS comes into play. Key in any staffing analysis is an understanding of both the tasks that must be performed and the need to know the consequences of them not performed, or performed on time. Discussions with console operators on key upsets and their required activities is the best way to understand the tasks and the impact of their omission. Some of the typical questions and how to interpret the operators' responses include:

- *What are you looking for?* The goal is to understand how the display system is utilized. If the operators talk about accessing a large number of displays or not having enough screens, then it is likely a sign of a poorly designed display system. Lack of an overview or higher level graphics are typical deficiencies in this area.

- *What actions are you going to take?* If the operator has to make a series of actions, then these become candidates for automation. For example, the standard response on one unit to water in the feed (fairly common) was to reduce rates. However, with numerous furnaces with eight passes each, a rate reduction encompassed a large number of control changes. This could be done automatically to a certain extent, relieving the operator of performing the action and freeing them to monitor its success.

- *What can go wrong?* The goal is to assess where operator intervention is needed to protect safety, the environment, or assets. If the operator sat on their hands during the upset, and the only impact was loss of production, then that is a risk that management can choose to take. If, however, failure to act results in a release or equipment failure, then some means to automate that response should be implemented.

- *If you had an extra operator, what would you have them do?* In many cases, the response to this question is "someone to sit in front of the alarm summary to filter alarms." Good alarm management practices should make this unnecessary.

- *What is the hardest part of handling the upset?* If the answer is in troubleshooting or managing the decisions and volume of information, then training is a likely point to ensure that this is not an issue. Display design and formatting may also be an issue here.

It may be that improvements to the OPS are not feasible in the short term. It may cost too much to automate some of the systems, or the display technology is too antiquated to make display improvements. In those cases, it may be necessary to have more console operators than required for steady-state operation. However, that should be a transient state of affairs.

5.2.2 Field Operator Workload and Staffing

"We don't staff for upsets" is the standard position ascribed to plant management where hourly personnel are employed. I have heard a few managers say it, although it is not as common as it is attributed to them. I maintain that the statement is false; plants do staff for upsets. They may not staff so that the upset has no impact (impossible for all upsets), but they do staff to safely bring the unit to a stable condition. If they didn't, then every time a unit had an upset, it would burn to the ground. (I did have one plant that maintained they weren't too far from that.) Plants do and should staff for upsets, particularly for the field positions.

It is assumed that additional personnel will be contacted and requested to come to the plant to assist, but they are not likely to arrive before at least 60 min. Therefore, the base staffing of a unit must be able to safely handle the key tasks associated with ensuring that the plant is stable while awaiting assistance from off-shift operators. Restart of a unit can be done when additional personnel arrive.

A basic premise is that if staffing levels are adequate for the most severe upset, then they are adequate for all less severe upsets. Therefore, the first step is to determine the most severe upset. Since the operators in an area can and do assist each other, the upset analyzed should be one that impacts the entire area. While an isolated upset might be more severe for a particular part of the plant, its isolated nature would allow assistance from other personnel, thereby distributing the workload over more operators. For example, a tube rupture in a fired heater is very serious, but it has limited impact on the other units. Operators from the non-affected areas can assist the operator on the unit with the tube rupture.

Plant upsets typically result in all available personnel responding for the duration of the upset. ("You can't have too many operators in an upset.") Therefore, direct time (time spent on job related tasks) or utilization is usually 100%, regardless of the number of personnel available (rarely is there a maximum staffing for an upset). If direct time is always 100%, how do you determine the minimum number required?

Upset workload is characterized by a sequence of tasks that must be completed in a relatively short time. If sufficient personnel are not available, then some tasks will not be completed in time. The consequence of failure to complete the tasks is what determines upset staffing requirements. While some consequences must be avoided (endangering safety, environment, assets), others may be deemed an acceptable risk by plant management (lost production, wasted materials, inefficient operation). While keeping the unit running through an upset may incur greater workload requirements than shutting down, being able to shut down the unit is a requirement, whereas keeping it running is a choice. With the exception of terrorists, it is universally acknowledged that being able to land a plane was more important than being able to get it into the air or keep it flying. With process plants, being able to safely shut it down is more important than being able to run it.

In our original upset analyses, we would format the tasks the operators had to perform into a flowchart to visually check for any conflicts. The problem was that it was largely qualitative in nature, confirming or denying that the staffing was adequate. It did allow identification of bottleneck tasks that drove staffing, but it didn't say how much was gained by automation of that task. We started to rate the tasks and consequences, mostly to identify those of high effort or significant consequence. Eventually, by playing with combinations of those metrics, we found that the product of the two seemed to capture the combined risk effort of the task. The only addition was a weighting factor to capture whether the task was a back-up to a board task (close a control valve), an automatic system (auto-start of spare), or backup for a safety instrumented system (close valve behind emergency isolation valve). Applying the ratings to a variety of units allowed us to establish a threshold for addition of operators. The result was the ability to quantify the upset workload and the impact of making tasks easier to perform or implementing automation.

An example of the results of this approach is shown in Table 5.2. Three different upsets were evaluated on the unit, with a loss of recycle (cooling medium) generating the highest workload and utilization. The unit was currently staffed with two operators, but the company desired to go to a single operator, which could occur if several tasks were automated, including remote shutdown of the charge pump and realignment of tower flows. Automation reduced the tasks that the operator must perform, and therefore, the overall workload required to respond to the upset.

Table 5.2.   Example of upset analysis

Upset	Operator utilization (Full time equivalents)
Loss of recycle	1.07
Loss of cooling water	0.93
Loss of power	0.85

Upsets determine how many operators must be present 24/7, but (hopefully) upsets are rare. What tasks are going to occupy the bulk of the operators' time? Will the tasks fully utilize the operators who must be present to respond to an upset? The answers to these questions require determination of steady-state workload and utilization.

We had two different clients who tried to list all the tasks that each operator position performed and had foremen estimate how long it took to perform the tasks. The results in both cases were operators who performed more minutes of work than there were minutes in a shift. When I challenged one of them on the apparent warping of space–time, he replied, "Oh yes, our guys are really busy." Not only did the results violate laws of physics, it defied their own assessment of walking in the control room and seeing operators at some point taking a rest. The results are not a surprise to human factors professionals, as people are very poor at time estimates.

The sure fire way to ensure accurate task time is to measure them, so we started using a stopwatch (now a handheld computer) to figure out how operators spent their time. Unlike an assembly line, the tasks for operators vary widely in when they need to be performed, fluctuate from day-to-day, and are under some discretion of the operator as to sequence. Our solution was to sample operator activities for long blocks of time. We were sure that the operator performed the recorded activities, because as we saw them do it. At issue was whether what we observed was representative of the job as a whole. Repeated samples would eventually converge on the average for the job. What we found was that the average field operator spends about 73% of their time on job-related tasks, but this average is a product of individual jobs that ranged from 19.75% to 100%. We target 80% for a fully utilized field operator. This is both consistent with the data we collected and the general target for personnel utilization in the human factors literature [23].

The steady-state utilization can be compared to the minimum staffing requirements to determine staffing options, as in Table 5.3. Utilization in the figure is the current workload divided by the target workload. In the case of Unit A, they required more operators to handle the routine work than was needed for upsets. Reduction in the routine work seemed unlikely in the near term, but a large percentage of that work occurred on days (maintenance).

Table 5.3. Example of staffing requirements

Unit	Required staffing	
	Steady state	Upset
A	3.32	2.34
B	5.03	5.70

Hence, they shifted personnel from nights to days to better balance the work-load with personnel. In the case of Unit B, their options were to either find additional work for their operators during routine operation or reduce the upset demands (and staffing) through automation improvements.

5.3   CONCLUSION

Operator workload, and therefore plant staffing, is neither a constant nor fixed with some set of equipment. It varies with the design of the other performance-shaping factors. For process plants, console staffing should be based upon the steady-state workload with the OPS designed to enable that staffing level to manage upsets if and when they occur (and they will). Minimum field operator staffing must be sufficient to safely bring plants to a stable condition in response to severe upsets. This will vary with the level of automation present.

6

Interface

I posed some questions about the displays they were using to an operator. I asked, "What do the purple lines mean?" (not sure if magenta would mean anything to him). He responded with, "Let me see. Those are infrequently used lines … no wait a minute; we use that line all the time. I'm not sure what the purple lines are for. You should ask Daryl. He built these."

When I talked to Daryl about a variety of things, I asked, "What do the magenta lines on the graphics mean?" He was pleased to discuss his work and he said, "Magenta is for mixed phase flow (combination of vapor and liquid). I used one color for vapor and one for liquid, so I needed one for mixed flow. I really wish there were more color options available, I ran out of them fast. Maybe the distributed control systems (DCS) manufacturer could create alternating patterns of color, like wires on telephones, such as dashes of green-brown, green-brown."

Daryl needed more color options like Bill Gates needs more money.

Interface design may be the most misunderstood aspect of human factors, or at least it is the most visible (pun intended). I am guessing that one reason for both the problems and some misconceptions is that anyone can create a visual display. Maybe not a good one, but anyone with a basic sense of drawing can likely sketch out what they think something should look like on a screen. In fact, at the beginning of more than one display project, the plant engineers have asked if it would help to give the operators sheets of paper to start sketching displays. (The answer then and now is no!)

Good display system design is more than what meets the eye, literally. What is shown on a screen is only a small part of a well-designed display. Good display systems, or what is nowadays referred to as "high performance HMI," are well organized and structured, with the content of each display selected and presented for its intended use. There are two facets of display system design: (i) content and (ii) presentation (or formatting). Of the two, content is far more important than presentation; if the content isn't present, no amount of fancy graphics tricks can compensate for it. However, if the display has the right content, and a poor presentation can hinder its use, at least the necessary information is present. The esthetics of a display is of minimal, if any, importance.

When I explained this to one client, he mentioned that while they had negative comments on formatting, they had never had a comment on content. My response was, "You likely never will; it is hard to realize that the information isn't present." The content and formatting become interwoven and bonded once the graphic is created. People assume the content is correct when given a display and focus on the presentation. Detecting poor content often requires more detailed analysis.

6.1 CONTENT

I observed a console operator at work in a refinery gas plant and noticed that a variable on his graphic had the engineering unit of "Wobbe." Since at the time I had never heard of that, I asked him, "What's a Wobbe?" The operator was equally stumped, so he called the head operator and repeated the question. The head operator promptly trended the point, and said, "I don't know what they are, but we seem to be making a lot of them." Having exhausted our collective knowledge on the subject, I let it drop. Later I found out the Wobbe number is a measure of the heat value of a gas stream per its specific gravity. I was told this was used in conducting heater efficiency calculations.

Why was there a point on the graphic that clearly had no value to the operator nor impacted his behavior in any way? The answer is because it was being measured. This is characteristic of a machine-centered approach to man–machine system design.

Don Norman, in his book, *Things That Make Us Smart* [24], likens display design to telling a story. I find that to be a useful analogy. A good story has been crafted, with the key points organized for maximum effect. Most displays in process plants are a stream of consciousness, without William Faulkner's writing skill. I suspect if you asked the display designers to tell their life story, it would be a single chronological listing of everything they remember, from their first Christmas presents to their last vacation, often in agonizing detail. Imagine looking at their graphics for 12 h or using them to control a volatile process during an upset ("Get to the point!").

The colors and shapes used in graphics are akin to the words and grammar of story. If a story doesn't have a point, then all the pretty prose or fancy vocabulary cannot make it interesting. A famous example is Noam Chomsky's[1] only entry in Bartlett's Familiar Quotations—

Colorless green ideas sleep furiously.

Clearly this is meaningless, but the grammar and vocabulary are correct. Likewise, a display system can use all the right colors and shapes but still fail to fulfill its purpose. You can follow all the guidelines for color use and symbols, but if the content is not correct, then performance will be hampered. Each display needs to make a point, and the displays need to combine into a coherent story.

Proper selection of content requires an understanding that just because something exists or is true does not make it relevant or necessary. Imagine you are teeing off at the golf hole in Figure 6.1. I am your caddy and begin to go into elaborate detail on the lay of the green, the pin position, the bunkers around the green, and so on. At some point, you would likely point out that the information is meaningless at this point, as you are a good three strokes from the green. I reply, "But what I said about the green is true. I thought you might like to know." "Not now!" would be your response. The value and need of information is not constant; it varies with the circumstances of its use.

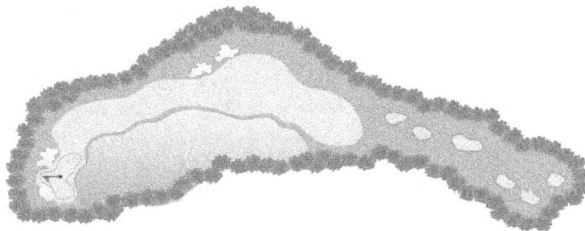

Figure 6.1.   Content Selection Example.
Source: ©GoGraph.com/adroach.

---

[1] Brilliant linguist, whacky political commentator.

6.2   STRUCTURE AND ORGANIZATION

I told a client that the operators were having a hard time navigating the system. The cause was a lack of structure. Our movement through virtual environments, like physical environments, is easier when the features are organized into a coherent structure. When I returned 3 weeks later, the operator who had designed the displays said, "I took your comments to heart. I worked an entire *week* on developing a map of the display system to make it easier for the operators to find displays." In my mind, I wondered why at day two or three into this effort it didn't occur to him that the problem was not the lack of a map, but a lack of well-structured system.

The structure for the information, the display system, should be hierarchical in nature. There are several reasons that point to the value of a hierarchy. First, the discussion on short-term memory showed that all the information is essentially captured in successive higher order chunks. Second, the work of Jens Rasmussen described earlier shows process control to have a hierarchy of behavior. Finally, recent research into the functioning of the neo-cortex, the area of the brain that gives us our intelligence, is essentially hierarchical in nature [25]. Only a few layers of cells are needed to organize the massive amount of data from our senses by using a heirarchy that captures sequences of patterns similar to that in Figure 6.2, with each level in the heirarchy a layer of cells in the neo cortex.

So the content of the display system should first be hierarchical. The general pattern is that shown in Figure 6.3. The pinnacle of the hierarchy is the over-

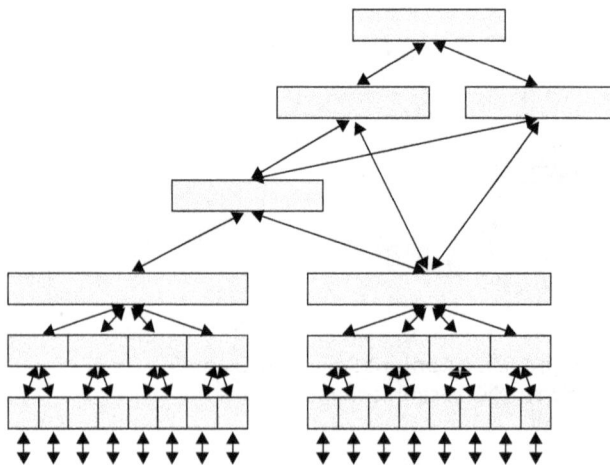

Figure 6.2.   Hierarchy of Neo-Cortex.

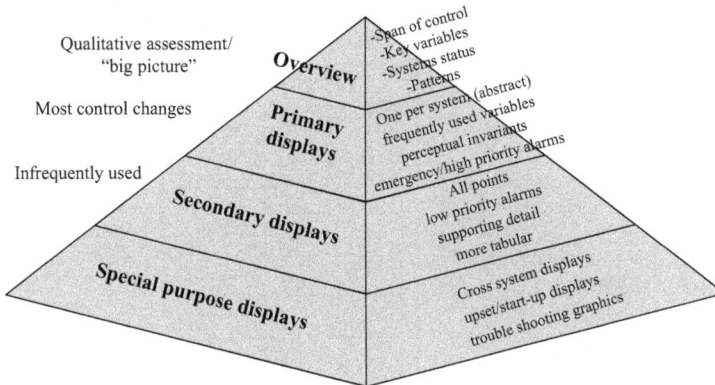

Figure 6.3. Display System Hierarchy.

view that provides the big picture of unit operation. It answers the question, "Do I have any problems in my span of control?" Below the overview in the hierarchy are the primary graphics, containing key process variables, critical controllers, and higher priority alarms. It is often said that 80% of the control actions should be able to be done at this level. That value is *not* to be taken literally. It is a way to describe the intent of these graphics and that most of the control changes should be concentrated onto a handful of displays. Next are the secondary graphics that contain all the points in the system. These are used when additional detail is needed to understand a problem or alarm. Finally, some special purpose displays may exist that either cut across functional boundaries or relate to special operations (e.g., start-up).

An example of how an overview is created begins with Figure 6.4. For this process, the operators thought of it in terms of five systems; within each of these sections, there were further breakdowns into multiple subsystems. This creates a structure for the information, that is an outline, for the display system story. It also shows how the overview is not independent of the system below it; it should reflect the structure of the underlying display system.

The specific process data needed for each subsystem is specified, as in Figure 6.5. You already would have noticed that no presentation formats have been offered; this state just defines the structure and content. Too often display designers will be influenced by what they "think" can fit on a display: "I can't get all of that plant/system/unit on one display." That is the totally wrong approach. The displays should be driven by what is needed, with formatting techniques used to accomplish it (make it fit).

The resulting overview display for this process is shown in Figure 6.6. When I first attempted to create an overview for this process, the designers felt it wouldn't fit. When the content of the display was specified, it also wouldn't fit according to designers. But clearly it fits, and without clutter.

Fire & gas		Emergency shutdown	
Inlet		**Deepcut**	**Utilities**
Slug catcher	Dehy	Gas-gas / Sales gas / Turbo / Absorber / Deeth	Fuel gas / Flare / Heating / Drains / Power
Deeth	Ovhd comp		

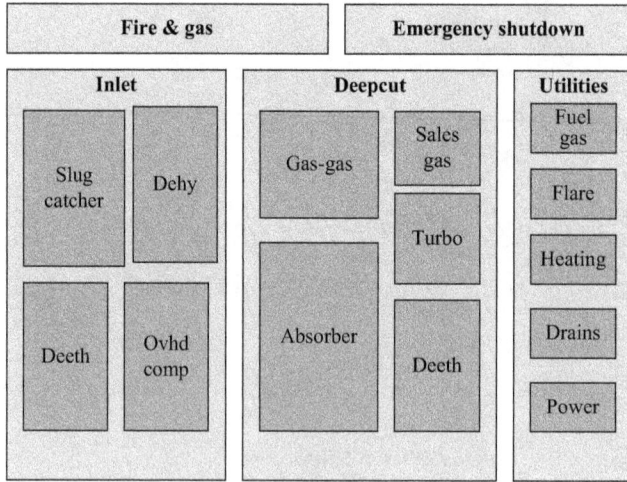

Figure 6.4.   Display System Structure by Systems and Subsystems.

FGS

	1300	1125	1100	1700	1120	1045	Admin	Util
FIRE								
GAS								
HEAT								
SMK								

SAFETY

	1300	1210	1035	1125	1100	1005	1990	1700	1120	1045
Fault										
Actuation										
Bypass										
Overide										

Time to plnt ESD ___ minutes   Time to Slug C ESD ___ minutes

GOLDBORO

CHANGE ZONE

INLET DEHY

Pig Receiver
press

Slug Catcher
press
level
flow gas out
flow liq out
level water boot

Inlet Deeth
level
flow product
pump status
flow feed
flow reflux
temp rblr cntrl
temp e1060 inlet
analy c1035 inlet

Ovhd Comp
press deeth ovhd
press deeth inlet
speed comp contrl

Dessicant Twr
analy dewpt
analy
analy
analy h20 content
temp out
press 3105 rcycl
dpress mercury twr
flow regen gas
temp regen htr out
flow regen gas comp
status regen blwr

level d1130
temp e1125 regen
dpress dehy filter

DEEP CUT

Gas/Gas Exchgrs
ratio of three matrls
limit cool down
dpress
analy expndr train
temp in exchgrs
temp out exchgrs
**flow feed e1175/95

Absorber
dpress e1205
level
temp e1205 hot out
temp e1195 cold out

Residue Comp
press dewpt vs deepcut
press d1180 inlet sep dewpt
press d1180 inlet sep deepcut
press k1300 comp dishc
press k1301 comp dishc
flow daily tlt
**flow sales gas
temp sales gas
analy sales gas moisture
analy sales gas dewpoint
speed
surge cntrls

Turbo Expander
flow expdr sep inlet
level sep lvl
temp sep inlet
temp drop dehy excgrs, jt, turbo
temp d1180 inlet
press d1180 inlet
press split range cntrl jt/turbo
**flow k1185 inlet
speed k1185 turbo speed ctrl
anti surge

Deeth
level
analy
temp liq tray 37
temp e1195 cold out
temp e1215 rflx cold out
temp e1195 hot out
flow reflux
flow pipeline pumps
   analy btu's
   % speed
   margin to surge
   engine speed
   flow inlet
Product Pipeline
temp pump
press pig launcher

UTILITIES

Fuel Gas
press d1600
press d1600
press d1600
temp e1610 htr out
levels scrubber
press d1620

Flare
level ko drum
mode pumps
temp ko drum

Power Gen
status gen 1700
status gen 1701
status gen 1725

Hot Oil
temp
flow min
flow to units
temp exchgr outlet

Glycol
temp htr outlet

Drains
temp
level

Figure 6.5.   The Specific Process Data Needed for Each Subsystem.

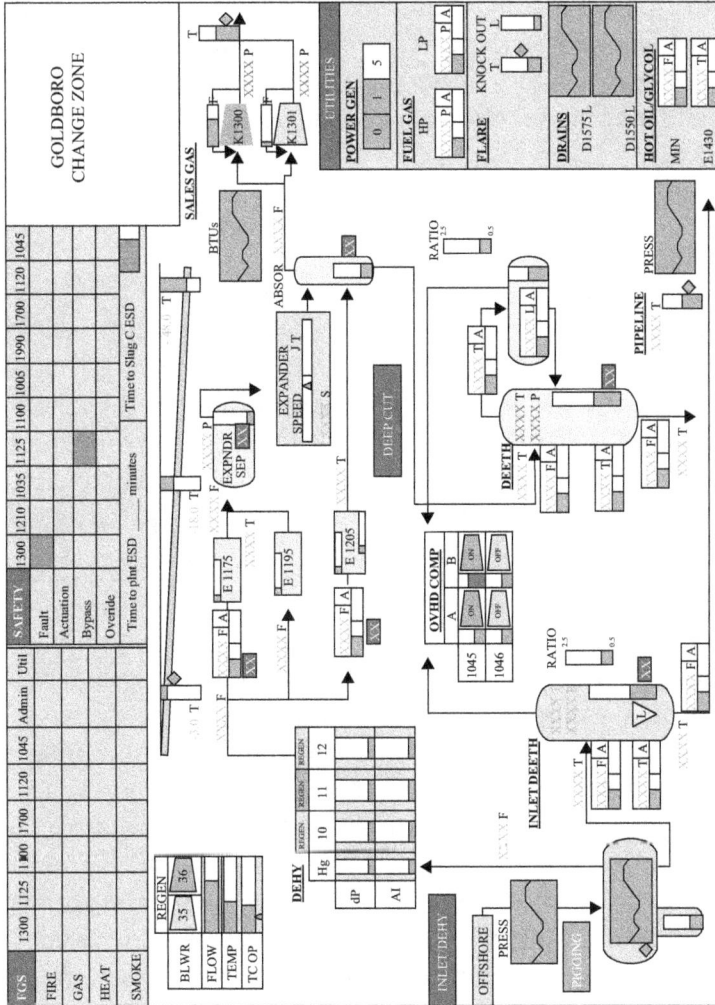

Figure 6.6.   Resulting Overview Display.

The overview display is just one display. I am amazed at locations that have multipage overviews. If you go beyond one display, it's no longer an overview. This is akin to a five-page abstract. The whole point is to condense the information so the assessment can be made from a single display.

There is an argument that more than one screen is available, so why not use it, but there is a cost for spreading out information. As Christopher Wickens noted on research done by Goldstein and Dorfman, "People can process high rates of information emanating from a single source more rapidly than lower rates emanating from several sources [26]." There is a value in a compact display system. Obviously, this needs to be balanced against other variables of clutter and chunking, but all else being equal, compact is better.

6.3   WORKSPACE

> A project engineer showed me the console design for a single operator position. There were over 10 DCS screens for a single operator. I asked, "Do you think 10 is the right number?" The engineer thought for a moment and replied, "I think 10 is enough." Clearly I had asked the wrong question, so I rephrased, "Don't you think 10 is too many?" He quickly replied, "No, I can spread the display system out to use them all." That's exactly what I was worried about.

In the last century, the workspace for a distributed control system was straightforward. There was a monitor with a keyboard, and one display per monitor. That is no longer the case. One keyboard can control multiple monitors. Each monitor can have several windows open at a time. How will these all fit together? How will the operator utilize the system? This needs to be thought through as part of the organization of the display system. Figure 6.7 shows an example of how the screens will be used. Each area has a specific defined use, either the graphic, controller pop-up, or alarm window. Some designers try studiously to avoid making any assumptions about the system to be used: "I want to provide maximum flexibility for the operator on how to use the system." The result is usually chaos, a system with little or no structure, imposing greater effort on the user to decipher how to interact with the system.

Along with greater flexibility also has come greater viewing area. In the early days of DCS, three screens were the norm, and additional workstations could cost over $50,000 each. Now, additional monitors can be bought at Wal-Mart, so the number of monitors for each operator has exploded. How much can a person look at? Human factors research has determined the

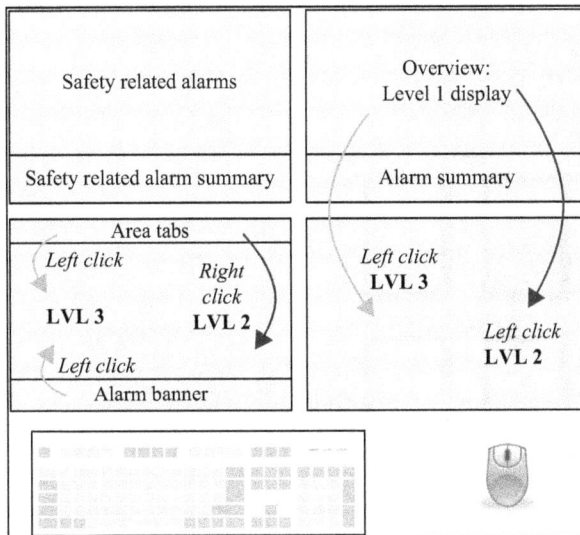

Figure 6.7. Screen Division Example.

optimum viewing area for a person, particularly where color is a key component of the displays [27]. Figure 6.8 shows the primary viewing area and how that compares to different monitor layouts. Since the area is a circle about 28 inches in diameter, the critical information should be confined to about two monitors, or four if stacked. The larger each monitor is, the fewer monitors that can be effectively viewed. Therefore, the designer should attempt to design a system that utilizes two to four monitors. Anything above that number is likely to have an exponentially diminishing value.

6.4 DISPLAY OBJECTS

One company had a standard for cyan text (light blue) on black background. There came the "revelation from high on" that gray background should be used, so they just changed the background color. Now they had light blue numbers on a light gray background. I almost went blind in the few hours I was sitting with the operators. In the name of "human factors" they had made their displays illegible.

The brick and mortar of display design are the objects that will be arranged on the screen. This is usually the shape library that will be used to implement the design concept (the content and structure in a defined workspace) that has

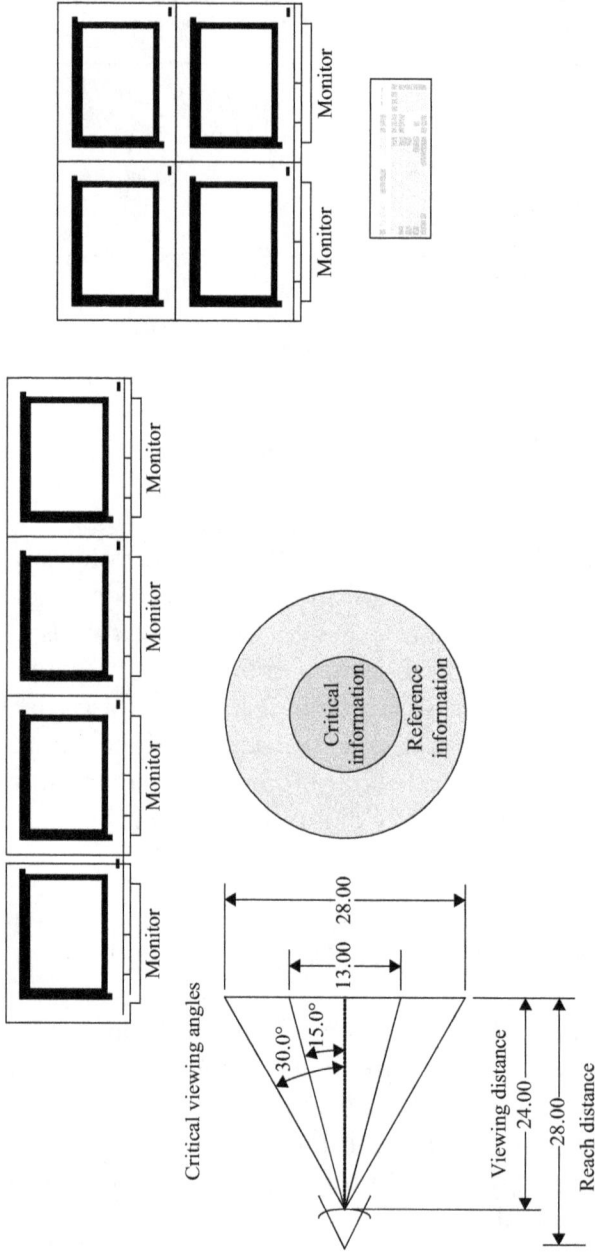

Figure 6.8.   Primary Viewing for Various Layouts.

already been specified. Two issues are often poorly understood and/or implemented: object size and the use of color.

Objects, including text, need to be sized to be legible or readable (there is a difference). This means that they need to cast a sufficiently large image on the retina. As the retina is curved, the size of the required image is in minutes of arc. The required size can be calculated using the formula in Figure 6.9. Generally, a visual arc of 12–20 min is recommended [28], and this applies to both objects and text. To maintain the same size on the retina, the object must be made larger as it moves away from the viewer.

Since objects need to be larger the farther they are from the viewer, the current move to use large monitors in a control room for overview information may not necessarily achieve the desired outcome. Just because the monitor is larger does not mean that more information can be displayed. Using 28 inches as the standard viewing distance, Figure 6.10 shows that a larger monitor might contain less data than a smaller one placed closer to the viewer.

One area of confusion we frequently encounter with object sizing is with regards to text. People ask what font size they should use. However, font is specifying pixels, not size, so the answer has changed with technology. The higher the resolution of the monitor, the more pixels there are per inch. This means the pixels are closer together. So a 12-point font will get progressively smaller as the resolution is increased. As a result, specification of the size of text should be in required height, not points of font.

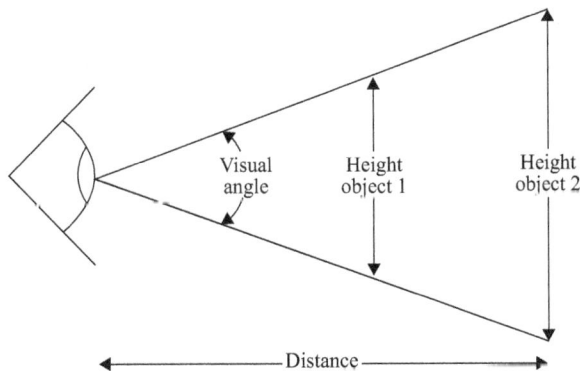

$$\text{Visual angle} = \frac{k \times \text{height}}{\text{distance}} \, , \; k = 3438$$

$$\frac{\text{Visual angle}}{k} = \frac{\text{height}}{\text{distance}}$$

Figure 6.9.   Visual Angle.

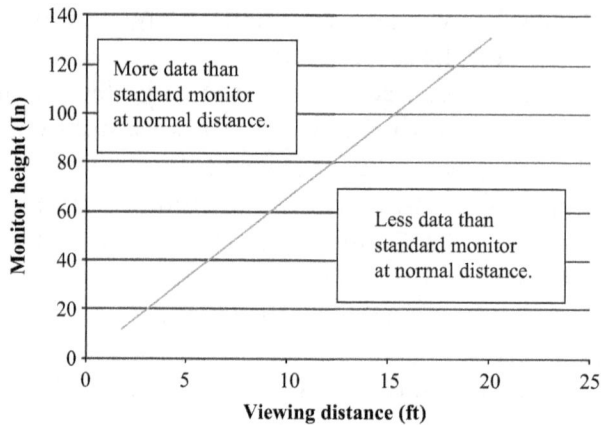

Figure 6.10.   Monitor Height Versus Viewing Distance.

Confounding the text sizing issue is the expected viewing distance. We did a project setting the text height using the standard dimensions for an operator sitting at a 90° angle with their hands on the keyboard. This is the 28 inches we used earlier. As anyone who has been in a number of control rooms would know, rarely are the operators sitting in such a manner. So the "real" viewing distance needs to be taken into account, perhaps the distance with the operator pushed back from the console, feet on the console being optional.

No current book on human factors in process control would be considered complete without addressing the use of color codes. The absence of color, or gray-scale graphics, is a major issue in display design. Before tackling the use of gray, it is worth understanding the overall objective in the use of color.

Color is just one of the major methods to encode information. The other three are location, alpha-numeric, and shape. The goal is twofold: one is to increase information transfer in terms of bits/in^2. Encoding takes less space. The other objective is to potentially enable parallel processing of information through the utilization of different cognitive channels. For either, there are certain principles of what makes a good code [29]:

- *Consistent*: If you are going to use a code, use it everywhere. This can be prob-lematic when the system comes pre-set with colors for certain items (for example, the setpoint or process variable) that may differ from how you would like to encode them.
- *Unique*: A code should have one and only one meaning. So red either means stop/closed or warning, not both. Consider the word "close." Does it mean "to shut" or "near"? Actually either, depending on context. It's ambiguous.

Contrast this with "stop." It's clear and unambiguous. Thus, a code needs to be unique in its use so as to be unambiguous.

- *Meaningful*: The purpose of a code is to transfer information. Therefore, that information should have some impact on behavior. The example of the mixed-phase flow code is a case of a code not being meaningful.
- *Detectable*: Again, since the purpose is to transfer information, the code needs to be detectable. One plant placed a minute "p" in the middle of a pump symbol tied to a seal pressure alarm. We hunted all over the graphic to find the pressure because no one could see the letter.

A good color code must meet the aforementioned objectives. You would have realized that the set of objectives does not mean that color should not be used, but that it be used for a unique purpose. In the past, I had to argue to use less color, but now I have to fight for companies to use more.

So, what about grayscale? Doesn't human factors research say that gray backgrounds are best? The basis for the gray background is often stated dogmatically as having come from a "study" that shows it reduces eye fatigue. Few of the people making that statement can even name the study. I have met only a few people who think they might have read some study along those lines. Without being able to refer to the study, it is impossible to determine how applicable the results are to a given circumstance. Who were the participants? Did they use process operators or Borneo cave dwellers? Were there any interactions or caveats in the results that need to be taken into consideration? How well did the study control extraneous variables? What variables were controlled for? How strong were the results?

Compounding the problem is that along with gray backgrounds have come generally monochrome displays, the foreground being different shades of gray. Therefore, is the problem the gray background or the lack of color? In order to tackle one problem or variable at a time, this section will focus on the background color. For reference, Figure 6.11 shows all possible pure colors on pure backgrounds plus gray [30].

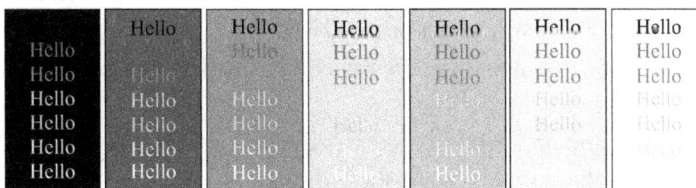

Figure 6.11.   Background and Foreground Color Combinations.

If eye strain is the issue, what causes eye strain? The major origin of eye strain is eye muscle fatigue. Viewing objects at less than 20 feet causes the eyes to converge, so anything closer than that will produce some eye strain. Any factor that causes someone to read at a shorter viewing distance will promote eye muscle fatigue and, therefore, eye strain. This is exactly what occurs when legibility is low; the reader brings the material closer to the eye, which effectively increases the font size (visual angle). Low achromatic contrast also decreases legibility and causes shorter viewing distance. A different set of muscles are used to bring the object in focus. The wavelengths that produce each color are imaged at different depths behind the lens. This means that we have to refocus the lens to see each color sharply. Saturated colors result in more refocusing. This constant refocusing can cause the muscles that change the shape of the lens to tire and may produce a sensation of visual fatigue [31].

However, eye strain is not the only issue. Research on legibility and pleasantness contradicts some of the research on eye strain. Eye strain research is generally concerned with reading from the display, a focused task over a period of time. So the choice of background depends in part on the objective: Is the goal to maximize legibility or to ease reading for long periods?

So what does the research say on background color? A brief inquiry revealed some interesting results. On the whole, *there was no consensus on optimum background color*. Two studies specifically recommend gray while three recommend some shade of blue. Much of the research dealt with light versus dark background, rather than gray background *per se*. The arguments and data for and against gray background are provided here for the reader to make their own decision. As you will see, often the research results are contradictory, which should lead you to realize there is no one "right" answer.

The argument for gray background includes the following:

- Black objects on white background are slightly more visible than are white objects on a black background [31]. However, black on medium gray and black on dark gray had significantly faster reaction times (i.e., best readability) than black on white [32].
- Gray background helps people keep their attention on the foreground text and graphics, reduces the likelihood of floating characters, and allows the designer to use the color black in the display [31].
- Performance was not different for color-coded versus black-and-white shape coding [33].
- An achromatic background increases the perceived differences in colors [34].
- Visual search with a color-coded display is enhanced with an achromatic background [34].
- Text readability depends on the ease of distinguishing letter and word shapes, which in turn depends on the discrimination of fine detail. High achromatic

contrast maximizes this aspect of perception. Since the chromatic system has 1/5 the spatial resolution of the achromatic system, color cannot produce fine detail. However, the reading rates for variations in brightness and color contrast were equal with large text. So, the high color contrasts that the gray background obtains can be mitigated by text size [30].

While it may not be gray, you may want a light background based upon the following:

- Black print on white background is more familiar because it simulates the text presentation in books and newspapers [30].
- Any pure color produces an after image, a sign of visual fatigue. This is true of pure red and green, but also of pure black-and-white. So any nonpure color background, of which gray is one, is better than a pure color such as black [31]. All light-colored text on dark backgrounds causes eye fatigue [35].
- The higher mean luminance level of light backgrounds produces higher viewer adaptation level, which produces better acuity. Reading smallish text depends on fine visual acuity [30].
- Higher luminance contrasts correlates with higher legibility, which correlates with higher pleasantness [36].
- The higher brightness level causes the pupil to shrink, resulting in two beneficial effects. First, it improves retinal image quality by reducing spherical aberration—misfocus due to small refractive errors in the lens and cornea. Second, as all photographers know, the smaller light aperture increases depth-of-field. This means that the print stays in focus over a greater range of viewing distances. This can greatly help some people who wear bifocals [30].
- There is much less screen reflection of glare sources with a white background. Try turning the computer off and note that you likely see reflections from other light sources in the room. The white screen washes out and often eliminates reflections. Of course, it is usually possible, and probably desirable, to light the room so that there is minimal glare and reflection from other light sources [30].
- If scanning between a dark screen and white paper text, the pupil must accommodate the difference in brightness [37].

However, there are reasons not to use gray, including the following:

- Some of the benefits of a light background only work if the surroundings are light also. A light background with dark surroundings can be just as bad as a dark background with light surroundings. The Illuminating Engineering Society (IES) in the United States recommends a maximum ratio of 3:1 for a visual task and the adjacent surroundings. White surfaces reflect about 80%

of the light and black 5%. We take these two percentages, divide 80 by 5 and we get a 16:1 light reflectance ratio if a light background is used in dark surroundings [38].

- However, dark text on light background was better under all ambient illumination conditions, with lower ambient light increasing legibility of projected displays [36].
- The color blue has a relaxing effect on the nervous system, and some studies have shown that it increases productivity when used as a background color. Blue is also a natural appetite suppressant [39].
- Blue should not be used for text [35]. However, black or blue text had the highest legibility [36].
- Color graphics were preferred over black-and-white in the ratio 7:1 [33].

So what does all of this mean? First, display backgrounds do not have to be gray. Second, display backgrounds probably should be light, but not white. Third, some shade of blue is likely a good background color.

6.5 LAYOUT

> A display designer was showing me what he had created. One item that stood out was that all the pumps were at the bottom of the page, resulting in complicated lines and numerous line-crosses. When I asked why the pumps were at the bottom, he said in all seriousness, "Pumps sit on the ground." How stupid of me not to know that!

Assuming that the content has been specified, the structure of the system established, and a library of objects created, then the last step in display design can be undertaken. This is the layout of the objects on a screen to create a display. Here the curse of piping and instrumentation diagrams (P&IDs) raises its ugly head, coupled with some odd concepts on how we use information. Before addressing the problem with P&IDs, it is worth discussing how the pieces of information placed upon a screen are processed by the user.

The creation of good displays is often the result of a misguided idea of how we process visual information, a paradigm of naive realism. This term was coined by Harvey Smallman to capture the tendency of people to think that making displays as close to reality as possible is the best approach to information transfer [40]. The fallacy in this approach is thinking that the brain is a screen upon which images are projected (Figure 6.12). The brain and central

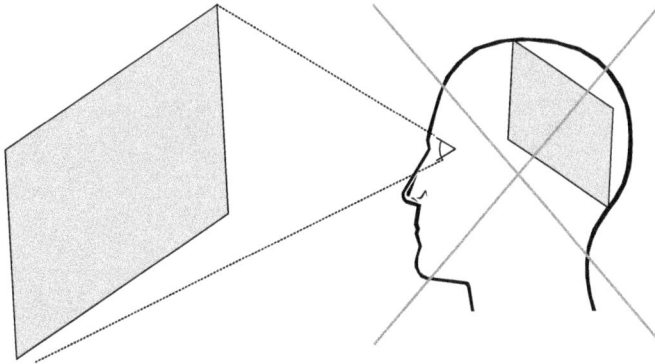

Figure 6.12.   Paradigm of Naive Realism.

nervous system are processing information as multiple points from the retina to the brain. Smallman's research demonstrated that while realistic displays were overwhelmingly favored by the user, performance was 70% better with more abstract displays.

What you think you see is not exactly what is really there. Your brain has learned how to interpret the inputs from the environment to create the appropriate image in your brain. How do you know that a smaller object is really smaller or just farther away? How do you know that a shadow on the wall is indeed that or just a different color of paint? When an object is rotated, why do you realize that the object has maintained its shape despite how the two-dimensional projection on your retina has changed? What we call optical illusions is based on taking advantage of the processing our brains do. Consider Figure 6.13; most people see a lighter triangle in the center, despite it having the exact same color properties as all the other white areas of the paper [41]. Your brain is not a screen upon which an image is projected; it is an information processor.

Learning that realism is not even necessary, or necessarily the best approach to presenting information is critical to creating truly useful displays. Taking another quote from Don Norman [42],

> The powers of cognition come from abstraction and representation: The ability to represent perceptions, experiences, and thoughts in some medium other than that in which they occurred, *abstracted* away from *irrelevant* details. (emphasis added)

Let's look at a very familiar object, which is in reality a highly abstract representation: a clock, shown in Figure 6.14. What is the clock depicting? Most people will say passage of time, but that is not the essence of the physical

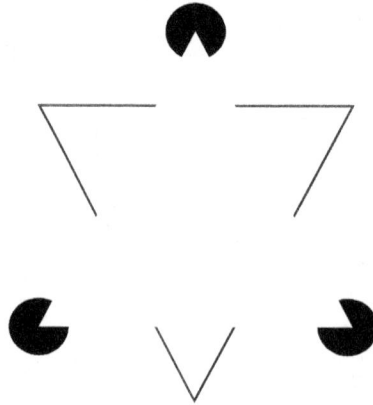

Figure 6.13.   Optical Illusion.

Figure 6.14.   What Is the Clock Depicting? *Source:* ©iStock.com/Pietus

phenomenon. A clock is depicting the rotation of the earth on its axis. It doesn't look like the earth, yet we can all tell "time" from it. It does its job in a very abstract way. Modern DCS designers would want to have a display of the earth turning for you so as to suggest time, some wanting to ensure that the continents looked exactly like they do. Our ability to process information is often better if that detail is left out.

Not only can we deal in the abstract, it is essential to optimize information transfer. Abstraction is, in part, a movement from quantitative to qualitative representation of the data. There is no right way to format data; it is only "right" in relation to its use, and qualitative representation is superior for certain tasks. In particular, qualitative presentation is superior if the task is qualitative in nature (e.g., good/bad). However, quantitative is superior if the task is quantitative in nature (e.g., "what is heater outlet temperature?"). A study by the U.S. Air Force using numeric values in a window display or vertical bar charts shows the impact of the presentation on the type of task performed (Table 6.1).[2] Task times in seconds are shown for solving the different tasks with different display types. I have included an example of quantitative versus qualitative presentation for a proportional-integral-derivative (PID) controller as reference, although this was not part of the original study. The U.S. Air Force found that significantly better performance time occurred when presentation matched the nature of the task.

Consider a system where the objective is to maintain a ratio of recycle flow twice that of the feed flow. Table 6.2 shows an example of that with four different representations. The top value is the ratio itself. While it is easy to determine if on ratio, the individual flows are lost. The second format uses different scales to maintain an easy to see ratio, gaining a little more detail on the flows themselves. The third format allows a further assessment of the individual flows, but the ratio is becoming harder to determine. The last example

Table 6.1. Impact of presentation on performance time

Display type	Time by task type		Example of quan/ qual distributed control systems (DCS) formats
	Quantitative	Qualitative	
Quantitative (Window)	102	115	SP 78.0 PCT OP 62 MANUAL
Qualitative (Vertical)	118	101	

---

[2] Air Force qualitative versus quantitative.

Table 6.2.   Representation options

| Recycle: feed | 2.12 |

| Feed Flow | 0 ———————— 4 |
| Recycle Flow | 0 ———————— 8 |

| Feed Flow | 0 ———————— 8 |
| Recycle Flow | 0 ———————— 8 |

| Feed Flow | 2.65 |
| Recycle Flow | 5.62 |

shows the individual flows quantitatively, making calculation of the ratios a mental chore. But those flows are known in great detail in contrast to the first value of just the ratio. Which representation is the best? There isn't one best, just better for different uses.

Abstraction is an essential technique to condense information. With abstraction, stripping away some of the detail results in a loss of data, but it is done to the benefit of compression of information. Consider three transformations along the continuum from quantitative to qualitative, shown in Figure 6.15. At the quantitative end is a reactor temp profile showing temperatures at various places in each bed of the reactor. This can be abstracted into a series of bar charts, showing if any go into alarm. The exact values are lost along with the physical orientation in the bed, but less space is taken to convey whether the temperature is in its range. The more extreme abstraction would show only if a point was in alarm. Further loss is the degree by which the points are in alarm or in what direction, but further compression is achieved. Again, there is not one that is best, just better for different uses.

How do you determine what is irrelevant detail that should be removed? As with most of human factors, the focus should be on the desired output of the individual, the operator subsystem. The question to ask is, "How does inclusion of this information change the decision?" If the answer is, "It doesn't," then don't include it. If the temptation is to say, "But it might be nice to know," then that is a signal that it's unnecessary. Knowledge is infinite, so you can always add something that would be "nice to know." However, what is needed are those data that alter behavior.

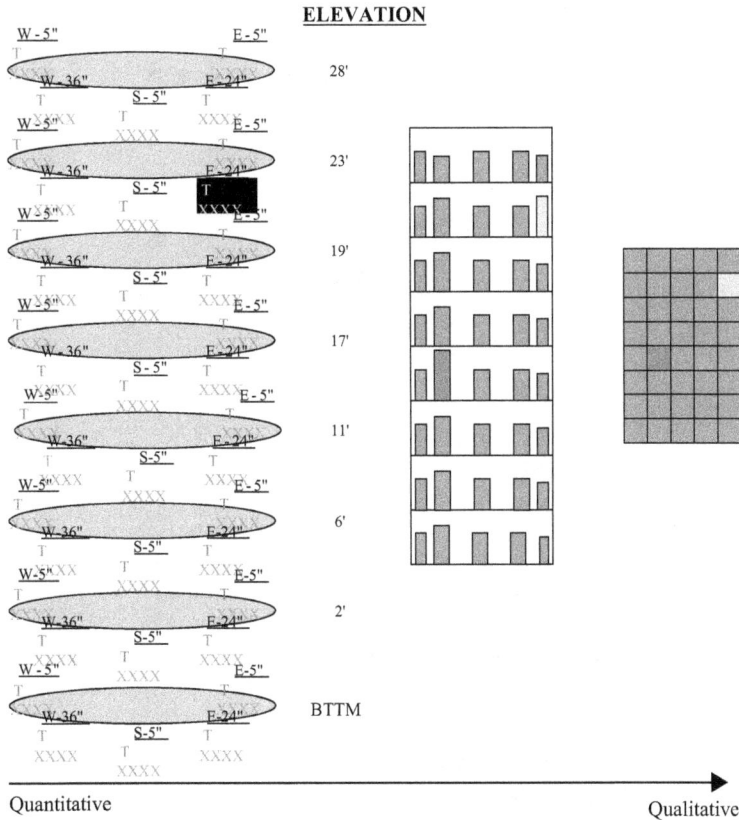

Figure 6.15.   Abstraction Examples.

There are some formatting errors that I see repeated on numerous displays. Some are well known, while others less so. There are a few general rules in formatting process displays that will help reduce the errors. These include:

- *Don't re-draw P&IDs*: The draftsmen who created the piping and instrumentation drawings for a unit likely at no time considered that these were intended to be used to control the process. Their objectives, constraints, and medium are entirely different from those of the display designer. Don't even use them for reference until the very end. One company sent their sketches of the displays they wanted to a third party company, along with the P&IDs in case there were questions. The preferred sketches were tossed and what they got back was the P&IDs configured as DCS displays.
- Figure 6.16 shows a portion of a display that attempted to recreate the P&ID and also what is possible if the focus is on just the critical information. The "after" version compresses the ten exchangers, which have no instrumentation

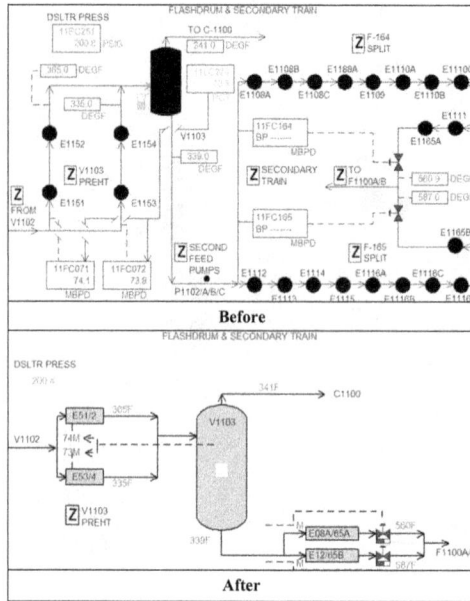

Figure 6.16.   Display Example.

between them, into a box listing the equipment numbers. What impacts the operator's behavior is that there is transfer of heat that increases the temperature of the feed, not how many exchangers are present.

- *Minimize noise by focusing on dynamic data*: Things that can change are likely to be the source of true information in a display. Static elements have the potential to stop providing context and increase visual noise. Visual noise is anything that is not a signal and does not supply actionable data. My favorite example is showing all the trays in a distillation tower. Unless the display has the ability to show the trays falling to the bottom on an extreme pressure spike, trays beyond those needed to identify draws/flows are visual noise. Static display elements are present only to provide context for the dynamic. The tower in Figure 6.17 has very little live data and a high level of detail on tower internals. Consider the information transfer of that object in bits/in². There is a large area with little information (three levels to be precise), so the information transfer is very low.

- *Enable perceptual invariants (patterns) to be created*: Where possible, attempt to arrange the information in a format that enables patterns to develop for the operator. The pattern need not be known in advance; a pattern to potentially emerge should be enabled. For example, aligning related flows, such as heater passes or feed, enable the operator to develop patterns that make changes and diagnostics easier. This is particularly effective if the flows are shown qualitatively, as in an analog bar. Figure 6.18 is for a unit that reacts differently to different feed sources, such that some require more hydrogen and/or result in

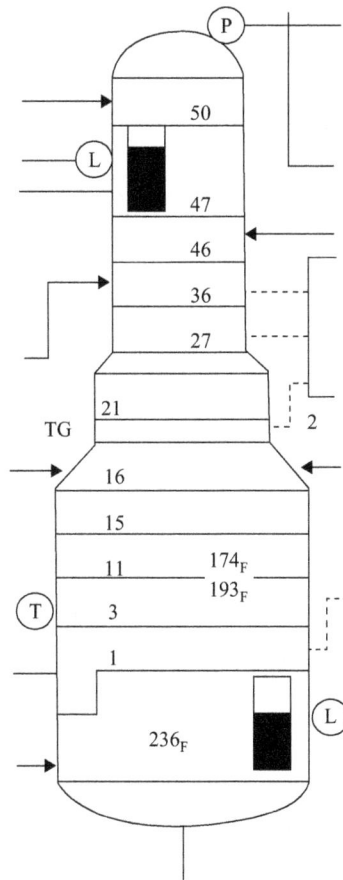

Figure 6.17.   Example of High Visual Noise.

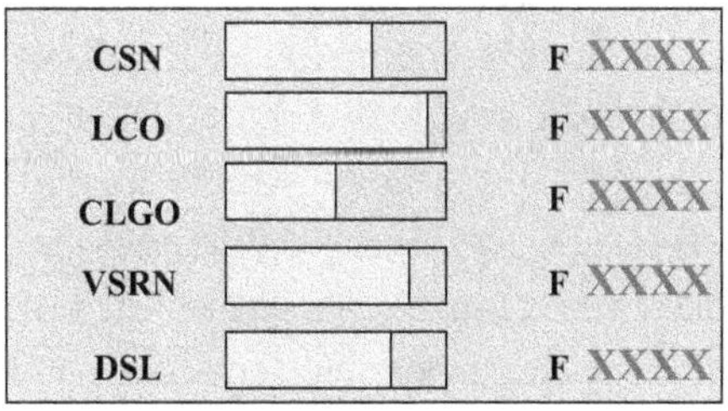

Figure 6.18.   Analog Bar.

high temperatures. The analog bars enable patterns to emerge from the figure, even if the patterns are not known *a priori*.

- *Match form to function*: If all the data is presented in the same format, it is likely that a mistake was made. This is generally because there are different uses for

Am I at setpoint?	What is my setpoint?
T	ATRC102 SP 500 PV 489.5 F OP 60 % MO AUTO

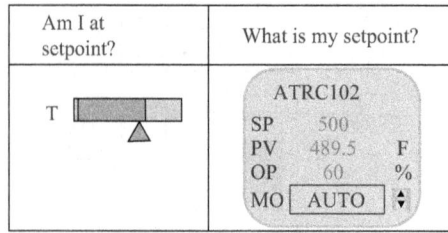

Figure 6.19. Formatting Options.

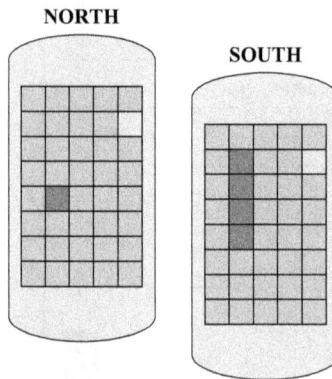

NORTH

SOUTH

Figure 6.20. Abstraction.

the data that will be better served with different formats. Some of the data will likely need to be presented qualitatively. Tasks that require comparisons are facilitated by tabular formats. Figure 6.19 shows different representations of the same data that best answer different questions.

- *Use abstraction*: Related to the first rule, the displays should employ abstract concepts. This may be as simple as where the flows enter a heater or as complex as merging exchangers into a single symbol with multiple designations shown (e.g., E101A/B/C/D/E). Figure 6.20 shows two reactors, no piping, and temperature indicators as qualitative status boxes of alarm/no alarm.

- *Demarcate/Enclose*: Where possible, enclose related information in either a box using demarcation lines or inside of the equipment to which it refers. Figure 6.21 shows variables enclosed by the heater to which they relate and free-standing. Enclosing the data facilitates visual search and eases short-term memory demands. Enclosure is the primary means to achieve high information density on a graphic without it appearing to be cluttered.

- *Careful about hidden items (absence of cues)*: There was a style in display design in the mid-1990s that made extensive use of invisible items, typically discrete alarms (switches). The intent was to reduce visual noise, which was worthwhile.

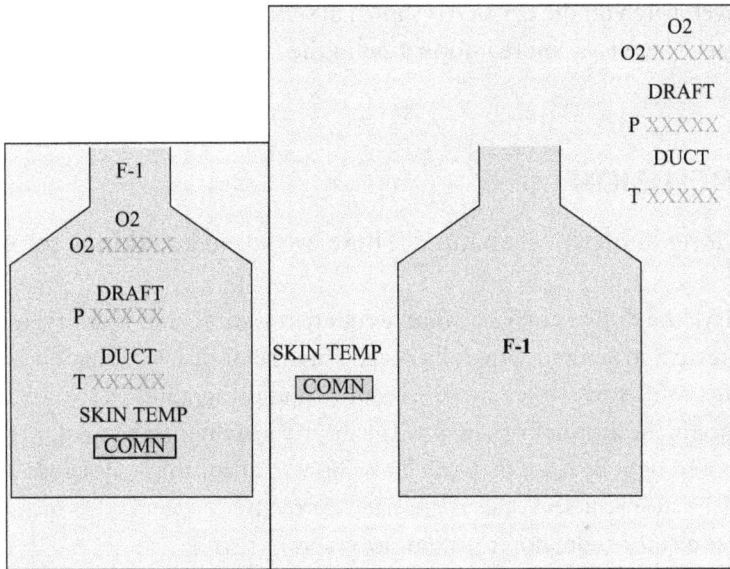

Figure 6.21.   Enclosing Data.

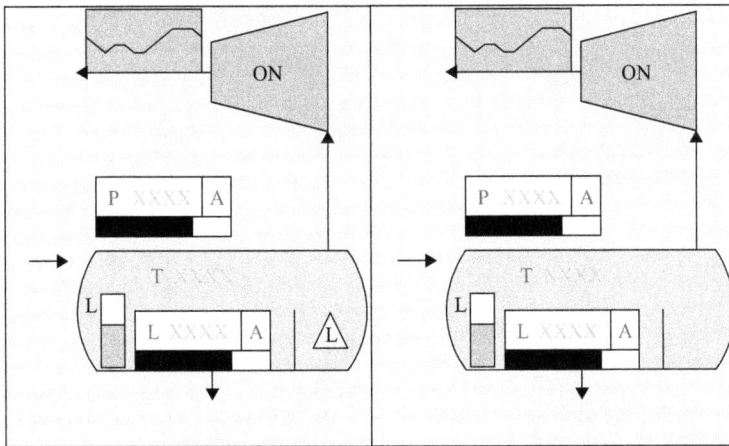

Figure 6.22.   Hidden Items.

The problem is that it also reduced information. People are not good at inferring something from its absence.[3] Seeing a discrete alarm that has not actuated conveys more information than a blank section of screen where the alarm is invisible. Figure 6.22 shows a suction vessel for a compressor. In the first, a

---

[3] Wickens absence of cues.

level switch on the dry side is shown in subdued color. In the second, it is invisible. There is far more information in the first than the second.

6.6   CONCLUSION

This was a long and critical chapter. There are some key things to remember.

1. Have the displays make a point, as content is more important than format.
2. Create a structured system by organizing the display information as you would organize a story: What are the chapters and paragraphs?
3. Specify the assumptions in how the display system is to be used, as the user will appreciate it because they will know how to adapt to the designer's intent.
4. Make sure font and objects are sized correctly.
5. Use a color code, don't just ignore color.
6. The user of the system is knowledgeable about it, and they will understand what is being shown even if it isn't identical to how it exists in the field.

7

Selection and Training

We were walking through the tasks an operator had to perform in response to a power failure in order to understand the tasks and the level of effort required. We briefed the operator in the control room as to what we were doing. Once outside, the operator asked, "What am I supposed to do?" I said, "Show us what you would do if there was a momentary power loss." "No," he explained, "What am I supposed to do in a power loss?" Clearly, he had no idea what to do. I suggested, "I'm not an operator on this unit, but securing the fired heaters is a typical first step." So off we went to secure the heaters. The operator had been working for less than a year but was considered fully trained and qualified. If I know more about running the unit than the operator, that's a bad sign.

I will qualify this chapter with the admission that we have not done much in the area of training, and whenever we did early on, it did not go smoothly. I admit that we became better as we learned our limitations and industry expectations. However, selection and training are key human factors concepts and we certainly encounter the product of both of these systems.

Any job requires skills and knowledge that are necessary for successful performance. Different individuals have varying degrees of different attributes that will either enable them to have the ability to acquire the skills/knowledge or influence the ease with which they learn it. Further, different individual attributes will impact an individual's satisfaction with a given position. Training is the mechanism by which the skills and knowledge for job success are imparted upon the individual. Selection is the mechanism to maximize the individual's potential for success at training and job satisfaction.

7.1   SELECTION

As we generally work for plant operations, all I know of the selection of operator candidates is that they undergo some testing and are chosen from a very large pool of applicants. The most recent practice is the requirement that the candidate has completed an associate's level program in plant operations. While this ensures that the student can handle science, technology, engineering and math (STEM) topics, it likely serves its largest benefit in providing candidates who have the desire and discipline to succeed.

In the last several decades, there has been a greater emphasis on use of a selection test for console operators. The rationale for this is that companies were finding too many console operators who did not have the requisite skill to manage the complete job. The flaw in this practice is that no one was failing the proficiency test for the job despite the company feeling that they had operators who could not handle the job. The main purpose of a selection test is to screen out those candidates who would likely fail the proficiency test, so as to not waste the students' or company's time. Since no one was failing the proficiency test, the use of a selection test to enter the console operator training program is not needed. However, rather than make proficiency tests that ensure the student can handle the job, companies have created selection tests to filter out individuals who will likely not be able to handle the job once they have passed the proficiency test.

The concept of a selection test is good. Choose those individuals with the attributes that best meet the job requirements. I have observed console operators for thousands of hours doing their job, with an opportunity to query them regarding how they did their job. I have developed a preliminary list of skills and attributes that seem to be present in the best of these individuals, as shown in Figure 7.1. Some of these correspond to the characteristics of expertise that Gary Klein has seen in other domains [43].

A recent finding in an unrelated field may have significance related to console operator attributes, which is spatial awareness. You may have done a spatial awareness test; they often have the form of a three-dimensional (3D) object unfolded into two dimensions. The participant must determine the shape of the 3D object from the two-dimensional parts. Interestingly, an individual's spatial ability influences the number of variables they think are important to performance, with those having high spatial aptitude requiring fewer variables than those with low spatial ability [44]. The latter tend to want to see variables

Console operators…….		
Are	Inquisitive	Investigate anomalies in data
	Decisive	Makes decision, doesn't become paralyzed with analysis
	Verbal	Both content and form of communication enhances information transfer. Ensures information was received. Being "chatty" is not the same thing
	Measured	Doesn't overreact to problems
	Cue/pattern oriented	Develops and responds to patterns or highly coupled process variables
	Thorough	Documents own and unit operational actions and changes
	Proactive	Heads off problems before they magnify
	Sensitive to context	Adjusts for changing contexts (e.g., rain, crew strengths) Have multiple plans for different contingencies
	Skeptical	Cross checks data, not trusting single parameters
Can	Multi-task	Is able to handle simultaneous tasks, balancing workload demands
	Visualize unit	Can mentally picture unit to enhance interaction/direction to outside operator
	Predict/extrapolate	Can predict ripple effect through the unit or impact from/to other units
	Maintain goal perspective (situation awareness)	Doesn't get caught-up in one part of the process. Understands impact of different process units on overall job
	Manage stress	Continues to function in high stress situations
	Take charge	Leads crew, articulating required response
	Tolerate boredom	Maintains vigilance over course of shift
	Plan/moderate workload	Can sequence various tasks over period of time to coincide with required unit activities, both for board and outside

Figure 7.1.   Skills and Attributes Present in the Best Operators.

that are not needed to solve the problem at hand. Why spatial ability impacts level of detail is unknown, but it seems to be true. One major oil company is looking at this to potentially supply different interfaces for users with different spatial abilities.

7.2   TRAINING PROGRAMS

The purpose of training is to ensure that an employee has the required skill, knowledge, and practical experience to perform correctly on the job. "To ensure" requires objective, measurable performance. "Required" implies management defined minimum requirements. "Perform correctly" necessitates an objective level of success or failure. From this, creation of a successful training program would seem to be straightforward, but the evidence is to the contrary.

I believe that all absolutes are false (For those of you who don't appreciate irony, that statement is ironic in that it is itself an absolute.) With that in mind, there are some common characteristics of the training programs that I have seen.

- *No one fails*: While this is slowly changing, rarely is someone hired who fails to qualify for the position. Console operators like to compare themselves to air traffic controllers. However, I have come across reputable reports that the failure rate for air traffic controllers was greater than 50% one time (I am told better selection tests have reduced that number.) Compare that with the failure rate for console operators, which I would guess to be less than a 10% failure rate.

- *Time to train is a constant*: The trainees' previous experience is irrelevant in how long it will take them to complete training, and every similar type of a job (i.e., board or field) takes the same length of time to learn. If you keep time to train constant, then students will leave training with different levels of skill/knowledge. Conversely, if the constant is the level of skill/knowledge, then time to train or acquire that skill/knowledge will vary.

- *Plant personnel think training is effective*: While many training departments acknowledge they can get better, most readily state that they do a good job. (Everyone thinks he is an above average driver.) This positive perception of training seems incongruent with the large number of incidents in which operator performance (or lack thereof) was an issue. Perhaps the fact that few fail training leads to a false confidence ("We must be good, no one fails").

Training problems are not unique to process industries. Walt Schneider of the University of Pittsburg identified common fallacies in imparting high-performance skills in a variety of settings [15].

- Fallacy 1—*Practice makes perfect*: Practice alone does not guarantee improvement. Often students lack the basic strategy/understanding necessary to create the input/output associations that practice reinforces.

- Fallacy 2—*Training of the total skill*: Teaching components of the total task can often be done faster and at lower cost than attempting to teach the total skill.

- Fallacy 3—*Skill learning is intrinsically enjoyable*: Too many programs fail to realize that students will get bored and miss the critical cues. Extrinsic motivators to raise attention are often needed.

- Fallacy 4—*Train for accurate performance*: Often the goal is to obtain acceptable performance on a specific skill, which will then break down at high workload when other tasks will be required. Overtraining or specialized task integration training is often needed.

- Fallacy 5—*Initial performance is a good predictor of trainee and training program success*: Initial performance of complex skills is highly unstable, often creating misconceptions about both the student and the program itself.

- Fallacy 6—*Once a learner has a conceptual understanding of the system, proficiency will develop in the operational setting*: This is the overreliance on classroom instruction with little guidance on how to use the information in performance of a task.

As you may have deduced, there appear to be few models of the optimal, functioning training program in process plants. However, the following are the elements that I look for in a complete training program.

7.3   SKILL/KNOWLEDGE REQUIREMENTS

As might be obvious, successful training requires specification of the behaviors that are desired on completion of training. Knowledge might be a lofty goal, but efficient training is directed toward creating the behaviors required for the job. What makes a good operator? This should be evident in a review of any training program, as its goal is to produce good operators.

The requirements for what makes a good operator are generally missing from most training programs. Some can at least infer what this is from the list of skill/knowledge requirements that is usually available. One plant calls this list their training guide, while others call it the training plan. Regardless, the list is rarely behavior-oriented, focusing instead on knowledge, with a heavy emphasis on process knowledge. An example of this approach is from a manual for a job shown in Figure 7.2. On inspection of the list, I found that the list did not include lock-out/tag-out and taking samples, two key tasks expected of an operator. The training coordinator's response was that the operators are supposed to know these tasks from their orientation training. Even if that is the case, it seemed odd to me that they don't have to demonstrate proficiency on the unit. With the exception of being able to draw the process equipment (I am not exactly sure how that helps on the job), there are few objective behaviors specified.

One of the baffling characteristics in training programs is the fixed time requirements. An individual who is new is seen as needing the same time to learn as an operator with 20 years of experience who has just transferred to a new unit. This defies any logic. More importantly, there is seldom opportunity, and certainly not incentive, to "test out" of some of the training requirements. While the total time to learn is specified, it is rarely delineated into a daily or weekly basis: What should the trainer have learned at the end of each day, each week? Table 7.1 shows a simple set of training topics for each day; the student and instructor have an idea of the pace they should attempt to achieve.

Training programs should attempt to build on skill/knowledge already learned. Using the assumptions of what the individual knows or has already learned, the training modules can focus on the additional level of skill/knowledge required of a particular position. This kind of modularized approach also aids in an individual in transferring or maintaining skill levels as they qualify on new positions. Once the fundamental or generic knowledge has been mastered, adding to it is simpler by virtue of how the new is different from the old.

```
Introduction
        Know purpose of heater, job duties, location of equipment, alarms, shutdowns
        Be able to draw equipment, startup/shutdown, and prep for maintenance
        Classroom (furnace, feeds, site visit, process flow diagrams (PFD))
        Making a round
        Training checklist
                Equipment location
                Learn to perform (trace lines/flows for 60 items)
Basis of design
        Process
        Firebox
        Draft
        Feed
        Sulfur injection
Detailed process description
        Pyrolosis
        Cracking heater
        Equipment design limits
        P&ID list
        Maintenance checklist
        Process sampling locations
Safety
        Fire extinguisher inspection
        Safety equipment inspection
        Pressure safety valve (PSV)
Environmental (one paragraph)
Utility systems (locations)
Instrumentation
        Alarms & shutdowns
        Solenoid operated valves
        Control valves
Standard operating procedures
```

Figure 7.2.   Example of Training Guide.

Table 7.1.   Example of Daily Training Requirements

Hour	Day				
	1	2	3	4	5
1	Surge	Pressed Oil	Ketone	Inert Gas	Stock Switch
2	Vessels	System	Tower		
3	Towers	Slack Was	Solvents	Sampling &	
4		System	Reflux	Logistics	
5					
6	Furnaces	Product Wax	Fuel Gas	Process	Circulation
7		System	Flare	Theory	
8	Stratco	Soft Wax	Sump/Slop		
9		System			

To summarize, a good training program starts with objective requirements. The desired behaviors should be specified along with the skill or knowledge required to achieve them. Expectations should be set for the order and timing of learning all the elements, with suggested daily and weekly milestones. A standard progression through the units should be utilized, which allows skills and knowledge obtained in one area to be expressly built upon in the other areas. For example, once distillation is mastered in the gas plant, only specific differences need to be discussed regarding the debutanizer in the Reformer. In this way, not only is training done more quickly, but it facilitates retention by integrating the skills/knowledge over a larger set of tasks and areas. Incentives and means to move as quickly through the program should be provided to capture different levels of expertise.

7.4 INSTRUCTION

Like display design, training is plagued by the problem that anyone can be a "trainer." One plant I visited had two major areas, wherein they recently went from one shift supervisor per area to one for both areas. The individuals who were no longer shift supervisors became trainers. I asked how they were chosen. "The shift supervisors with the worst communication skills were made the trainers" was the response I received.

I have given training in display design. One student asked me to review his work a year later, stating he had taken my training to heart. What he produced was horrible (You got that from what I said?). I have sympathy for the challenges in being an effective trainer.

On-the-job training is the norm for most plants. The trainer is often just the senior operator on-shift; although some companies utilize mentors (operators who have received some training on instructional techniques), a few locations have dedicated trainers for a given unit. On-the-Job Training (OJT) is considered a useful form of training, as it places the student in the operational setting. The problem is that simply being in the operational setting is not a guarantee of learning.

With the exception of where dedicated trainers are used, the general instruction process seems to through osmosis. The premise is that if the individual is exposed to the skill set, he will eventually absorb it. Obviously this form of training has a history of some degree of success. However, I think the potential pitfalls are equally obvious. First, the skill must be demonstrated at some point in order for it to be absorbed. If the event that requires the skill never occurs in the course of the training period, the skill cannot be obtained. This is particularly true with console operators who may experience abnormal situations infrequently. Second, considerable variation can exist in how the skill is taught between different crews/individuals. Senior operators often say they can identify who trained someone by how they perform certain tasks. While some of the variation may be stylistic, some can be substantive. One process unit required a change on hydrogen line-ups in response to an upset. One crew did it one way (short loop) and the other a different way (long loop). Since both worked, it wasn't the option that was at issue. The concern was an operator learning from one crew and then working on a different crew. The key to successful team performance is to have everyone work from the same script [45]. Clearly, multiple scripts existed at this plant.

While having dedicated trainers might be good, it is, in many cases, impractical. The way to minimize variance due to instructors is through a structured training plan with objective measures of performance. A plan that both defines what the student is to cover each day, along with those skills that are to be demonstrated/practiced with the trainer, will at least ensure a more consistent presentation.

One oddity in process plant training is the notion that training occurs on overtime. I have spent many an early evening in a control room where there would be an hour or so of dead time. I have never heard of a plant where there is an attempt to cross-train operators in these off-hours. Even if the attempt is not to qualify the individual on another unit, there would be value in them understanding something about how the other unit functions. But apparently people only learn at time and one-half.

7.4.1 Training Material

Even when the training requirements are specified, that information is not always in the training manual. For example, a training requirement was, "Don't leave the unit until you have made proper relief." However, the characteristics of proper relief are not defined in the training manual.

Training manuals focus heavily on describing the process, with little reference to actions or behavior the trainee should exhibit regarding it. Here is an example from one manual: "If any liquid accumulates in the knock-out drum, it is pressured batch-wise to the hydrotreater product stripper overhead accumulator." It is my belief that the *trainee*, as the operator, will be responsible for initiation of the batch, and yet the manual almost implies that it occurs independently of human action. The manual fails to answer the questions such as: How is it pressured to the accumulator? Are there actions that need to be taken? Where? Any coordination with the console required? What can go wrong?

Where considerable resources have been devoted to "operator training," the result is usually a wall of ring binders. One operations manager said that his bookcase of training manuals was a total waste and could be thrown away. The plant's poor training material was given to a company that made it look glossy, but it was still poor training material (Garage in = Garbage out). Consider the

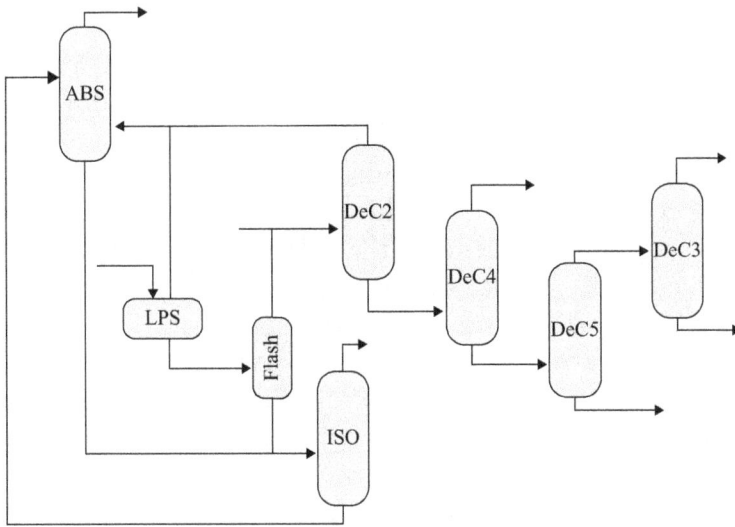

Figure 7.3. Simple Distillation Process, Complex Training Manual.

system in Figure 7.3. It consists of eight vessels engaged in simple distillation. The training manual for this was 488 pages long.

One of the causes of heavy training manuals is the penchant for bulking those manuals up with extraneous material, similar to a high school student's term paper. I had a high school English teacher who would say the grade for a term paper was proportional to its weight. Procedures, policies, drawings, and so on, are added to the training manual for no clear reason. Reference material, especially that which can change (e.g., procedures), should be separate from the training manual itself. The training manual should contain the instructions and explanations needed to master a specific behavior set, organized into daily segments.

7.4.2   Training Tools/Systems

Simulators are often seen as a panacea for console training problems. Many process plant managers are enamored with the high-tech training devices seen in aerospace. Many a simulator has been purchased only to later gather dust in some room. One simulator was purchased for hundreds of thousands of dollars, never to be used and later cannibalized for spare parts. The problem is not in the training tool, but in the lack of understanding by plants on why the tool is needed and how to use it. A hand grenade is an effective weapon, but to a caveman it's just a funny looking rock. If you don't know how to pull the pin, you won't get the full effect. (Or, if you pull the pin without knowing the function, it will have an adverse effect.)

Effective use of simulation requires an understanding of what skill/ knowledge is to be transferred. Consider our earlier model of operator–process interaction in which three key tasks are required of the operator: (i) detection, (ii) identification, and (iii) compensation. The primary breakdown for humans is in the first two. Most full-scope simulators like those purchased in process plants emphasize the third, enabling the operator to make control actions with realistic process dynamics. Fluency at the components of a task is necessary to master the whole task. Practice on the cognitive aspects of the job generally show greater benefits for improved performance than does practice on the motor skills (e.g., control valve adjustments) [46]. As automation increases, the motor skills become increasingly less important. Why use a human to control a variable to a target when the DCS can do it better? Leave the operator to set the target, making the decisions that we as humans excel at over machines. Simulations should focus more on the decision-making aspect of the job than on the physical output.

While many process company employees envy or attempt to emulate the aerospace industry in the use of simulators, they may not truly understand how simulation is used in that domain. There are at least 10 primary characteristics of aerospace simulator training programs [47].

1. *Functional context training*: Instead of attempting to teach each individual skill (e.g., making descending turns in an aircraft as part of landing), the skills are brought together in a larger context of a mission (e.g., instrument approach).
2. *Individualization of training*: All aspects of training are adapted to the learning rate of each student.
3. *Sequencing of instruction*: Training that builds upon previous training.
4. *Minimize overtraining*: Once performance proficiency has been obtained, training is ended (Note: This is one of the fallacies listed earlier).
5. *Effective utilization of personnel*: Instructors are matched to their talent areas so as to not have them doing work for which they are overqualified.
6. *Use of incentive awards*: Behavior modification techniques are utilized to motivate both the trainee and the trainer, such as more time off if training can be completed sooner.
7. *Crew training*: The trainer may alternate in their role (pilot versus co-pilot) so as to teach different facets of each job to the trainee.
8. *Peer training*: Using trainees to teach parts to other trainees.
9. *Minimize equipment costs*: Training tasks are allocated to the level of fidelity necessary to maximize learning.
10. *Objective performance measurement*: Objective performance measures so that assurance of proficiency is independent of who is doing the assessment.

How can or should the detection and identification skills be imparted? One technique that has shown promise is the use of decision making exercises (DMX). Developed for training decision making in combat by the U.S. Department of Defense, the technique was adapted to process control operators through a project sponsored by the Center for Operator Performance. The underlying premise is that decision making is a skill like any other, and practice is required to obtain or maintain proficiency at it. Rather than give operators "what if" drills wherein the problem is known, it is better to give them symptoms and force them to identify the nature of the problem. Add into the session time-pressure, distractions, and stress (have the operations manager watch but not comment), and decision-making skills can be honed.

I was part of the initial DMX development program for the Center for Operator Performance (COP). The problem I was given was regarding the shutdown of a pipeline pumping station for maintenance. I started another pumping state to maintain system pressures/flows and then shut down the prescribed station. The instructor told me that I just got a loss of communication alarm with the station, and asked if I was concerned. I said, "No, I just shut it down. I would be concerned if it were running, but not now. Maybe I won't be able to start the station back-up when maintenance is done." Next I was informed that a different station had tripped, and the trap had been sprung. The station I had shut down was the lynchpin in the whole pipeline. Without it, I had few options should another station trip (which was the scenario). And, since I lost communication and couldn't restart the pump station, the pipeline had to be shut down. As a novice, I had approached the problem in a very superficial manner. An expert would have understood that the loss of the station severely constrained operation.

I was the guinea pig in a second exercise that involved water in the feed for a fluid catalytic cracker (FCC). The scenario had swings in the flow controllers on the fired heater along with pressure spikes. I focused on the flow swings and went down the path of possible false surge drum level indication, which resulted in pump cavitation. During the debriefing, when I was asked about the pressure spikes, I said I didn't know what to do with them, so I ignored the data. Again, as the novice, my mental model was not developed enough to handle all the inputs.

I believe that achieving one of the biggest benefits to improving operator performance is possible, but politically volatile. For console operators, it would be possible to mine the control system data to quantify operator performance. This would identify those operators who perform the best, which most unit supervisors feel they can identify. These results could be used to assess both selection and training: What are the characteristics of the best operators? Did the best operators have training that others didn't? Did performance on the process improve after training or use of a simulator? The political part of this is that in addition to identification of the best operators, it would also identify the worst. So, while it might be done with the best of intentions—how to raise all operators to the level of the best, it also creates the potential to be used as an ax by simply cutting those that are at the low level of performance. Or, if an incident occurs with the lowest performing operator on the console, is the company liable in some way? However, ultimately selection and training need to be tied to process performance, a crucial element of success.

7.5   CONCLUSION

The element of objectivity is lacking in most operator selection and training programs. The lack of objectivity is true in all elements of training—from the specification of the attributes and skills an operator needs to the measurement of the effectiveness of the training. The latter includes not just successful performance in training, but success on the job as a result of training. This is referred to as transfer of training, an area that generally is not addressed in process plants.

8

Job and Organizational Design

We were leading a team that was helping to set a vision for the future at the plant. I asked for areas that the organization did well and those that could be improved. One member felt that the organization did not take criticism very well. When the individual was a no-show at the next meeting, I found out he had been reprimanded by his supervisor for making that comment. Apparently, they also lacked a sense of irony.

In this chapter, we move from the individual to the organization that creates the context in which the work is performed. The organization determines how the work that must be performed is allocated among individuals. The structure of the organization impacts the flow of information, particularly targets and constraints. Finally, the organization is responsible for the coordination among the various individuals performing those tasks.

8.1 TASKS TO BE PERFORMED

While the nature of the system generally determines what tasks need to be performed, it is management that determines who is to do them. For plant operation, Beville Engineering has identified—from observing over 1,000 operating positions across North America—five major tasks operators perform. These tasks and the subtasks for console and field operators are shown in Figure 8.1. Field operators have different subtasks, but the five primary tasks remain the same.

Task	Common	Console	Field

Operational	Implement day instructions
	Plan for contingencies in operation
	Meet production and quality targets
	Troubleshoot operations
	   Determine cause
	   Work with engineers
	   Provide audit trail
	Track unit condition (bypasses open, spares down)
	Assist with unit test runs
	Respond to upsets
	   Equipment failures
	   Process disturbances
		Optimize process operation
		Take action to correct alarm condition
		Investigate/resolve advanced control constraints
		Communicate changes with potential impact to another units
		Communicate process issues with management/engineering
		Adjust process variables to match operating plan/plant targets
		Communicate with/direct field operators
		Analyze unit operation (SPC/ Material Balance)
			Adjust process
			   Know product specs
			   Know key process variables
			   Communicate with board
			Shutdown unit
			   Emergency
			   Normal operation
			Take direction from board or supervisors
			Make correct line-ups
			Backflush coolers
			Swap pumps as needed
			Clean pump screens/filters
			Control material releases
			Lubricate equipment as needed

Inspection	Analyze equipment main functions	Detect and investigate anomalous process data
		Detect degrading process Conditions
		Match mid and long term path of unit to targets
		Ensure process resources (raw material, storage) are available
			Perform diagnostic tests and audits
			   Vibrations/bearing
			   Temperatures
			   Steam traps
			   Drain plugs and caps
			   Bonding and grounding
			   Critical alarms
			   Mechanical Integrity
			Physically check unit (leaks/ line-ups/ rotating equipment)
			Monitor and document fugitive emission
			Ensure unit labeling is complete

Administrative	Prepare turnover
	Participate in training (OJT/CBT)
	Prepare training material
	Track/Schedule training
	Train personnel (Operators/ Technical/Administrative)
	Identify training needs
	Participate in personnel requirements (time sheets/ arrange overtime coverage/ performance reviews)
	Read and revise documents (plant drawings/procedures/manuals/ management of change)
	Write task Safety analyses
	Participate in safety drills/ meetings/audits/inspections
	Investigate document near misses/injuries
		Perform shift relief
		   Review inhibited/disabled alarms
		   Review standing alarms and correct
		   Review controls out of normal mode and return
		Document unit Operation
		   Routing
		   Products/specs
		   Verbal orders
		   Process/equipment issues
		   HS&E
		Participate in HAZOPs/ rationalizations
			Relieve operator
			Log operational changes
			Maintain area
			   Order supplies
			   Keep area clean
			     Hoses rolled up
			     No trash
			     Tools stored
			     Clean up spills/leaks

Maintenance	Write work orders Assist with maintenance planning Prepare permits Audit unit maintenance backlog	Evaluate potential impact of maintenance activities on process Troubleshoot control system problems and document	Prepare equipment Follow lock-out procedures Assist with maintenance setup Monitor maintenance Ensure tool cribs are stocked Perform minor maintenance
Lab/quality	Understand and interpret lab results	Adjust process variables in response to lab results	Prepare sample tickets/bottles Pull samples Test samples (color, pH, gravity) Document laboratory results

Figure 8.1.   Five Major Task Areas that Console Operators Perform.

At one time, a single position was expected to perform some aspect of all of these functions (Figure 8.2). However, as Adam Smith noted in the late eighteenth century and early efforts in military applications of human factors discovered, specialization has some distinct advantages. Rather than have four individuals perform a portion of all the tasks, better performance can be obtained by allocating certain tasks to those individuals who can perform them the best. At the beginning of my career, it was not uncommon to have

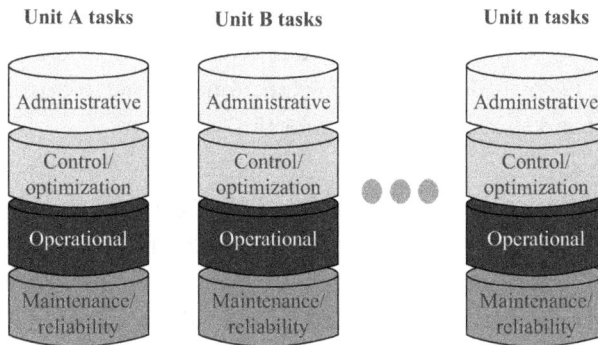

Figure 8.2.   Combining Tasks into Different Jobs.

operators who both made control changes on the control panel and prepped equipment for maintenance in the field.

One of the key areas where this specialization has manifested itself is with the use of console operators. The causes of the movement to having consoles as a specialized position include:

1. Distributed control systems (DCS)—A higher demand is placed on operator recall with DCS than with analog control panels. With analog panels the operator could scan all their instruments in the hope of spotting the anomaly. In DCS, the operator must choose what they want to look at. They need to have an idea of where the anomaly is located in order to find it rather than just stumbling across it as in the days of analog control panels. The difference is analogous to a multiple choice test (analog control) and an essay test (DCS). Different cognitive skills are needed to handle the DCS, which makes people with some attributes better at it than others.

2. Advanced control—Increasing advanced control means decreasing operator intervention in unit control (which is the objective), meaning less opportunity to maintain control skills. Degradation of skills has been experienced on many DCS systems, requiring that the operators start with superior skill levels and be away from the job less.

3. Consolidation—With consolidation of control centers, the board operator is removed from many of the traditional support structures (e.g., other operators). A more autonomous and independent individual is needed to handle the stress of these situations than in a conventional control room.

4. Increasing span of control—Control center consolidation made possible by DCS allows operator span of control to be increased. Often this increases the ratio of outside operator to board operators, with the net effect of longer periods between working the board if all operators rotate through the board position. A large span of control also further taxes the mental skills of the operator.

5. Emphasis on optimization—Plant management is increasingly asking the board operator to be more of an optimizer than a monitor. Quasi-control engineer skills are needed to fulfill this new function. Dedicated board operators enable selection and specialized training for this area.

While console operators were the first to become specialized, the trend has not stopped with them. An increase in operators dedicated to the maintenance/reliability function is occurring. With increases in safety shutdown systems, often more operators are needed on day-shifts than 24/7. The result is creation of operator positions that involve the same schedule as maintenance, with the focus on increasing maintenance efficiency.

8.2   JOB ROTATION

> Operators in a unit with four positions were complaining about having to learn and stay qualified on four jobs. I wondered how they would feel if the company cut one of the positions, thereby making them required to learn only three jobs. I had a feeling that wouldn't have gone down very well with them.

With few exceptions, operators are expected to learn more than one position. There seems to be a trend for operators to only have to know two positions, although why two was chosen is unclear. I have seen plants where operators are expected to know seven. Regardless, there is often a concern about having to learn "too many" jobs.

How many jobs are too many? How many jobs can a person learn? The problem with both of these questions is that there is no standard definition of what is a job or position. At one plant, there was a position responsible for overseeing six firewater pumps, all located in the same small building. That was a "job." I have also seen an operator responsible for eight cracking furnaces with waste heat boilers on an olefin plant. That also was a "job." Without a standard for what constitutes a full job or position, the question of what is too many becomes meaningless.

The issue, of course, is really about what can be learned and retained. The fear is that skill/knowledge will degrade if too many jobs need to be learned or performed. It is therefore necessary to understand under what conditions skill/knowledge degrades. Learning theory holds that there are three phases in the loss and recapture of skills/knowledge: (i) initial learning, (ii) decay, and (iii) relearning (refresher) as in Figure 8.3. Of interest here is the rate

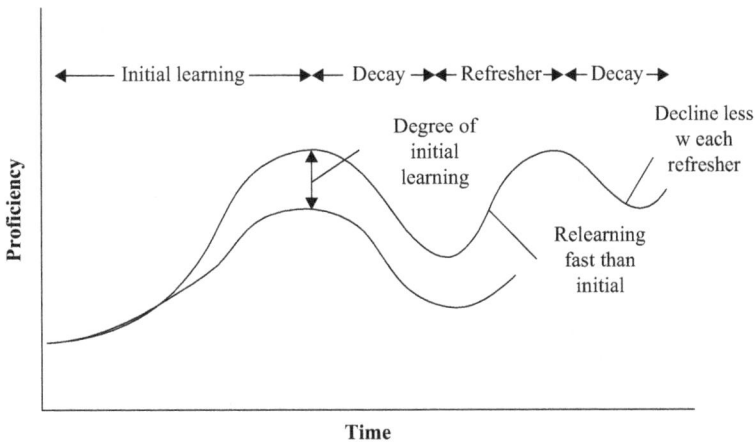

Figure 8.3.   Three Phases in the Loss and Recapture of Skills/ Knowledge.

and degree of decay. Generally, decay does not occur for about 4 months and whatever decay that will occur is complete within 9 months. However, the decay is never so severe as to take the learner back to their initial point. The amount of decay is also less after each refresher training session. More importantly, the rate of decay is a function of how well the initial material was learned. The more an individual overlearns the material, the slower the rate of decay. The rate of decay will also vary with the nature of the task. If the task has cues to help trigger learned material, then the decay will be slower. Also, if elements of the task are practiced as part of another task or even through mental rehearsal, then the decay will be less. Finally, relearning is faster than the initial learning, by about 50% [48].

To put it simply, no problem is anticipated as long as 4 months do not pass before an individual had worked on a job. While this would seem not hard to execute, I have encountered several facilities where individuals had not worked a position on which they were "qualified" for over 9 months. However, the skill/knowledge degradation will be slowed if similar skill/knowledge is utilized on a different job. If I take samples on both Jobs A and B, then the process of taking samples on one job generally refreshes the skill for the other. This same refresh function of skill/knowledge similarity should be built into the training itself. Once a skill has been mastered, it will take less time to learn it in a new domain. Training should build upon the base of skills that are acquired from previous training or experience.

The number of positions an operator must learn and the structure of the shift schedule have important implications for fatigue management. The American Petroleum Institute (API) has recently updated their recommended standards for fatigue management, RP 755 [49]. Under the new guidelines, work sets shall not exceed seven consecutive day or night shifts during normal operation (14 for turnarounds). To permit two consecutive nights of sleep after a work set, there shall be: a minimum of 36 h off, with a minimum of 48 h after a work set containing four or more night shifts, or a minimum of 48 h off after a total of 84 h worked regardless of day or night shift. While this seems easy to achieve, it has in fact proven problematic for many plants. Several have either been forced to violate the standard or operate short-staffed for 4 h of a shift.

While fatigue is often associated with high overtime, RP 755 has implications apart from overtime. Application of the RP 755 guidelines to a standard crew schedule shows considerable constraints in personnel usage (Figure 8.4). Only the white cells are days in which there are vacancies that can be easily filled. So for over half the month, there could be an issue or inability to cover a vacant job with the new guidelines.

The use of more day (maintenance) operators and operators qualified on multiple positions eases the impact of API 755. Maintenance operators, who generally work weekdays, can easily slide into a 24/7 position to cover for a daytime vacancy. If given sufficient notice, they can cover a night shift with little impact. The more positions on which each operator is qualified increases the flexibility in finding an individual to cover a vacancy. In contrast, having operators required to learn only one position, such as through job preference arrangements, risks skill decay and an inability to cover positions for unexpected vacancies.

So how many jobs should an operator expected to be qualified on? The answer would be they should qualify on as many as possible provided they work on the job periodically within a 6-month period. This increases the flexibility in scheduling and minimizes the potential to either violate API 755 or run the shift short staffed. The more similar the skill sets in the jobs the better, as it will be an inherent refresher on the other jobs. If objective levels of proficiency are built into the training program, it should be easy to determine if and how much skill degradation has occurred and whether the individual has retained the necessary skill knowledge to do the job. Obviously, these factors are intertwined.

Crew	M	T	W	T	F	S	S	M	T	W	T	F	S	S	M	T	W	T	F	S	S	M	T	W	T	F	S	S
A	N				D	D	D		N	N	N			D	D	D	D									N	N	N
B			N	N	N	N			D	D	D		N	N	N			D	D	D	D							
C	D	D	D	D				N	N	N	N			D	D	D		N	N	N								
D		N	N	N			D	D	D	D				N	N	N	N			D	D	D						

■ Unavailable
Can only cover day vacancy
Can only cover night vacancy
May be unavailable depending on previous shifts worked

Figure 8.4.   Impact of RP 755 on Crew Availability.

8.3 TEAM PERFORMANCE

A fluid catalytic cracking unit had just lost a critical piece of rotating equipment (main air blower). The console operator was frantically trying to shut down and stabilize the unit. Field operators were equally focused on troubleshooting the trip, shutting down pumps, and isolating/diverting streams. A call came across the radio, "Gas Plant Operator to Console." A terse and concerned reply came back, "What is it?!" A casual Southern drawl sounded, "All the samples are to the lab." That particular operator was clueless that a major upset was in progress, and instead set about delivering the scheduled product samples to the lab.

Everyone on a team can perform their job adequately, and still the team can fail. Team performance and teamwork are often emphasized in safety posters and pre-shift meetings, but exactly what constitutes teamwork is usually left vaguely explained. This is also observed in plant training, where the focus is on individual skills/knowledge, not team skills. However, failure of teams to function properly is at the heart of many "operator error" events. Successful team performance has been thought to be an emergent feature, where the whole is greater than the sum of the parts [50]. Two crews may look similar to each other in personnel, but the performance of the team will emerge when needed. Any sports fan has seen the better group of players bested by the better team.

So what makes a good team? Team performance and coordination has characteristic requirements for success and reasons for failure. Expert teams (as opposed to teams of experts) generally exhibit all or most of the following characteristics [22]:

- *Shared mental model/script*: Key to a team coordinating is that they are all "working on the same page." This is where preplanning of and practice for team response is essential, ensuring that the team knows both the common goal and how it is to be achieved. With good mental models, the members know what to anticipate from other members and can predict what will be occurring next. This reduces the need for communication and its associated workload. If scripts are practiced extensively, some of the tasks can be performed automatically, again resulting in almost no demands on workload for those tasks.
- *Monitoring performance*: Good teams have members who not only monitor their own performance ("I am not getting this done, I need help"), but team performance as well ("This isn't going right"). It has been noted that experts become more concerned with the process than do novices, who tend to be focused on the outcome. However, if the outcome becomes unattainable, the novice may not be able to recognize it.

- *Triggering conditions:* As part of the shared script, but also developed over time, is the tendency of expert teams to recognize the cues that initiate key tasks or serve as markers for performance.
- *Adaptation:* The reverse of the value of planning is the ability of the team to adapt to a novel situation. While training and planning to create the shared script is essential, it can't or shouldn't be followed at all costs. Good teams can recognize when their script is no longer applicable and a new ad-hoc script has to be developed. However, the more scripts possessed in common by the team, the easier it is for a good script to be developed and embraced quickly.

Surprisingly, teams can falter not just from not possessing the positive traits, but also from the nature of the problem as well. If the task requires extremely precise sequencing of tasks (little room for error), then teams can fail if the team is focused on efficiency. Tightly coupled task situations often require a more modular and less efficient team operation. It might be necessary to have one team member be in an idle/wait mode (seemingly inefficient) so that he can perform his task(s) at the correct moment. We see this with product switches, whether pipeline or coke drum, where an operator is given an advance notice to prepare for the switch and then wait till it occurs.

The nature of a team incurs a coordination cost, with bigger teams incurring a greater cost. At some point, the value of the team can be overwhelmed by the need to coordinate. I observed this in a plant that had split the span of control over a large number of console operators. One unit sent feed to another, which sent part of it back and some onto another, which sent some back to the first. These three tightly coupled units and their operators had to function like a bob-sled team. Their strategy in an upset was that one operator would yell out what he was going to do, and the others reacted. High demands for coordinated response offset any advantage gained by the console operators having less workload from a smaller span of control. Sometimes it is easier for one person to perform both tasks than having two people attempt to coordinate their actions.

An aspect of team performance that has gained attention in the last 25 years is crew resource management (CRM). Originally developed for the aviation community, the concepts have been extended to other domains. In many respects, CRM attempts to explain why team performance breaks down or fails to emerge. The key dimensions of CRM are shown below [51].

1. *Communication:* This is not whether it exists, but how is it performed. More specifically, is the communication loop closed such that the sender knows the receiver has received the message. I know of a plant where a foreman was rattling off instructions during an upset into a radio whose battery had died.
2. *Briefing:* This is not giving a briefing, but creating a plan to accomplish a given task. This is particularly critical for novel situations.

3. *Back-up behavior*: Do the team members know what is expected of them by the other members? Can they anticipate what their teammates need and provide it?

4. *Mutual performance monitoring*: What is not going right needs to be recognized and it should be voiced by any team member; the ability to follow that the script no longer applies.

5. *Team leadership*: Obvious in a team situation, but are team leaders chosen for and trained on leadership skills and abilities?

6. *Decision making*: Can each team member handle the mental processing demands of their position and make the correct decision at the correct time?

7. *Task-related assertiveness*: There have been several cases in aviation of a co-pilot following his pilot into a mountain or river even when he suspected pilot was wrong. Can the team members hold their ground until convinced that they are incorrect?

8. *Team adaptability*: We do not design for double jeopardy, but most major incidents have had more than a single malfunction. Can the team adapt to something not working or going as planned?

9. *Shared situation awareness (SA)*: While this is a major topic in itself, the CRM component is: Does the team have SA? This is an extension of the monitoring characteristic mentioned earlier.

Eduardo Salas, one of the developers of CRM, is well known for conducting unbiased appraisals of CRM's success. Despite overwhelming positive comments from those who had received the training, they couldn't find objective measures of success [52]. That does not mean that CRM doesn't work, but its benefits have proven to be elusive.

8.4 CONSOLIDATED CONTROL ROOMS

"This consolidation was a dumb idea" was the statement from the chief operator of a fluid catalytic cracker (FCC) unit. "What consolidation?" I asked, pointing out there was only one console and one unit in the control room. "The crude unit is on the other side of that glass partition," replied the chief operator. The glass was completely covered with calendars, safety posters, and memos so that I couldn't see the console on the other side. I asked if the consolidated control center was built with a glass partition. The chief operator explained that it was not done originally "but the crude unit operators work inside/outside, so there were times when everyone was outside and their alarms were driving us nuts, so they put up the wall." Well, something was a dumb idea.

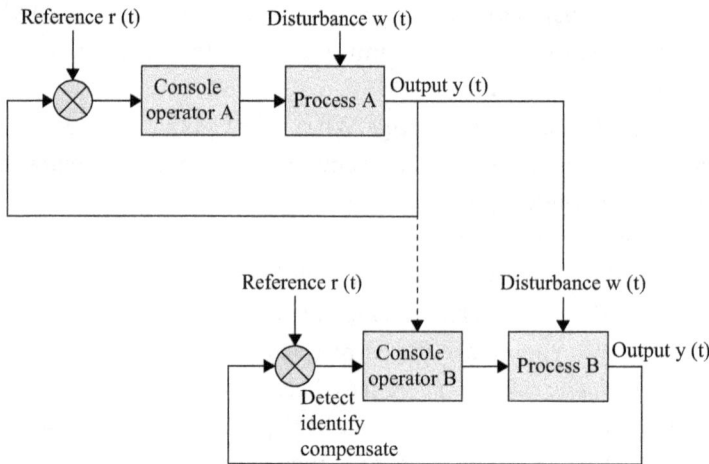

Figure 8.5.   Feedback Model of Operator–Process Interaction.

A specific case of team coordination is seen in plants with interactive process units. Utilizing a variation on our feedback model of operator–process interaction (Figure 8.5), we can see the impact when one unit is supplying material to another. If the output of Process A forms the disturbance to Process B, then the Process B operator still needs to detect and identify the nature of the disturbance. Enhancing the communication between the two operators enables the Plant B operator to respond faster; in certain circumstances, it might be possible for the Process B operator to respond prior to the effect of the disturbance being seen in the process. Almost as important, easy and effective communication between the two operators enables numerous small, incremental changes to be made rather than a few gross changes. However, communication is difficult if the two individuals are in separate locations.

There are several methods in which the communication between the two operators can be enhanced. One method is to place the two operators in close proximity. This is the basis for consolidation of control centers seen at many refineries. I was at one plant where the upstream unit called on the phone to ask if the downstream unit could handle an additional 500 barrels of feed. The operator at the front end of the unit then asked the same question to the operator responsible for the back end, and the response was relayed back to the upstream unit. This back and forth of telephone calls were repeated several more times over the course of the few hours I was there. It was admirable that the upstream unit even bothered to call, but the clumsiness of the interaction highlights the value of these operators being together.

A second method to enhance the communication would be to have all the communication occur within one individual's head, that is, have the same person control both processes. Many people intentionally separate the control of interactive units as an upset on one will likely cause an upset on the other. While true, it must be balanced against the potential performance gains that can occur when the same individual can quickly make the adjustments needed for both units. Consider the example above. Not only would having all the operators in the same control room be of benefit, but rather than having to make a verbal request from the front-end operator to the back-end operator, an operator whose span of control encompasses all the units could simply make the desired change. Additionally, they could likely make the change sooner and in smaller increments with an evaluation of each change.

If done well, consolidation of control centers can result in a step change improvement in plant operation. The benefits accrue from two sources: (i) those changes that tap into the change in system dynamics with operators of interactive units/systems in close proximity and (ii) utilization of the disruption caused by the consolidation to enact changes that might be difficult to justify or implement in isolation. The benefits are seen in four major areas outlined below.

8.4.1   Unit Operations

Improvement of plant operation with interactive units/systems is the chief benefit and justification of control center consolidation. While unit-to-unit interactions are usually the primary beneficiary, interactions between units and common resources should not be discounted. By co-locating operators, changes in one unit or system can be quickly and accurately conveyed to affected units. This enables more rapid changes to minimize impact on the plant. In decentralized control rooms, requests from other units for changes in operation were usually met with, "If they don't call twice, it must not be that important." Contrast this with a central control room where a simple increase in pressure on the flare knock-out drum resulted in a console operator yelling across the control room, "Who is venting?" Operators checked their console and the problem was quickly resolved. This particular location has reached the point where time in the flare is a fraction of what it was pre-consolidation. Improved plant performance due to the enhanced response to changes results in:

- Faster and more accurate response to upsets
- Better management of common resources (feed, utilities)
- Increased unit integration
- Optimization of the supply chain
- Reduced inventory
- Less off-spec material

8.4.2   Personnel Utilization

Better utilization of personnel is obtained from the opportunity created by co-location and the momentum of change created by the consolidation process. Console responsibilities can be realigned to take advantage of a single point of control for highly interactive units/systems. At the same time, the potential change in the organization can enable better utilization of both console and field operators. Jobs and job duties can be realigned to take advantage of fractional workload need or capacity in board or field jobs. For example, two one-half person jobs in remote control centers becomes a single job in the consolidated control center.

8.4.3   Work Practices

The momentum of change from consolidation creates a vehicle to alter work practices that would be potentially difficult to implement under normal circumstances. The consolidation creates a definable event that can be used to signify a new way of doing business. Some of the changes that plants have implemented include:

- New tasks/duties for field operators for better utilization of their time
- Higher skill level requirements for console operators reflecting their increased span of control
- Increased expectations for how business is to be conducted (a console operator turned control engineer/cowboy/chef at one location noted that they had raised the expectation bar to the point it was now a tripping hazard)
- New and expanded training to support the above
- Altered shift schedules to better manage fatigue.

8.4.4   Organization

Changes in the organization may be a necessity of the consolidation, but they can also be done to correct other issues that hinder optimal performance. The necessity is usually driven by distance, where a central control room may make if difficult or impossible for the supervisor in a decentralized facility to manage the same personnel where the console operator can be up to several miles away. A general rule of thumb is that decisions should be made as close as possible to where they are implemented, which the co-location of operators along with other key stakeholders can facilitate. At one facility, I observed a visit by the scheduler responsible for a product to discuss the coming day's plans with the console operator whose unit made it. No supervisors were involved, but just the two responsible parties. The net result can eliminate what had been bottlenecks in decision making, where decisions would move up the organization and stall. The product scheduler and console operator were located in the administration building, reducing the barriers for interaction between the console and other departments. Opportunities of increased interaction with technical personnel can yield similar benefits, allowing problems, particularly generic ones, to surface and be resolved faster.

8.4.5   Consolidation Failure

Simply putting operators into a common control center will not achieve the above. I know of several locations for which consolidation was a massive failure, to the point it was undone and decentralization reestablished. Other locations have stayed with the consolidation, but feel they did not see the promised or anticipated benefits. The failure of consolidation is usually either due to problems in the vision for the consolidation, poor implementation of the vision, or both.

A focus on geography often prevents plants from developing a vision of integrated operations with a central control room. The geographic paradigm is a mindset that says the plant should be organized based upon geographic proximity. This produces three possible errors in the vision. First, units that have no interaction might be co-located, thereby degrading the board-field interaction with no gain in board–board interaction. Second, the existing organization might be retained, despite having a potentially remote control room based on function/interactions. Excessive supervisor travel time or awkward lines of reporting may occur. Third, the geographic paradigm often results in the perceived need for the control center to be near the units. As the number of console operators in the control room increases, the need to minimize visitors increases as well, or chaos ensues. At some point, field operators will not be allowed to go to the central control room due to crowd control, making its location near the unit irrelevant and thereby enabling the control room to be placed anywhere.

The relocation paradigm is where management views control room consolidation as a simple relocation of the console operators. This is often done in the name of minimizing change and disruption. While it may do that, it also limits and potentially negates any benefits consolidation could bring. The most glaring problem from this approach is a poor span of control for the console operators. The number of console operators is set by the previous limits of pneumatic tubing in a digital workplace, with bored operators (pun intended) being the result. Almost as damaging, the same expectations and work that was done previously is being done now. There is no change in culture that the consolidation process enables and can facilitate.

A good idea can be rendered ineffective by poor implementation. Most of us have been on one side or the other of a cooking recipe gone badly. While we may politely blame the recipe, too often we know the problem is with the cook. There are at least five major errors that can ruin the consolidation process.

1. Poor design of the control room can quickly torpedo the consolidation. Control rooms that are too noisy, too crowded, or too isolated have all resulted in negative assessments of what should have been a good consolidation.

2. Poor communication systems. The very nature of a remote control room requires good communication. One plant had a single radio channel, resulting in an overload during upsets and operators being forced to frequently check in at the local control room for instructions. This worked as long as the local control rooms were "local." It would have been potentially disastrous if a remote central control room had been built. I observed a poor operator at one location holding a handset to each ear, as different units he was controlling were on different radio channels.

3. Failure to manage the transition. Central control rooms both necessitate and enable major changes in the operating culture. Those changes need to be managed or a major backlash can occur. The greater the degree of change, the more effort is required to manage it; but it is usually worth it in the long run as the opportunity will not present itself again for a long time. We discovered with one major oil company a need to have a transition manager, someone distinct from the project and day-to-day operations who ensured that the non-hardware changes that were part of the consolidation were achieved, including expectations, cross-training, new skills, and supervision [53].

4. Poor interface design, either alarms or displays, is often attributed to the move to central control. New versions of the distributed control system often precipitate or accompany movement to central control. If the interface is poorly done, it will often be blamed on the consolidation, with the bizarre notion that deconsolidation will correct the poor interface design.

8.5 CONCLUSION

Management has an impact on the performance of process operators through setting expectations, the design of their positions, and creation of a system that facilitates performance. The last item on the list requires an understanding that plant performance is more than the sum of individual operator performance and a product of the ability of the individuals to act in common toward a common objective. Consolidation of control room operators has the potential to significantly improve plant performance by both altering how interactive processes are controlled and enabling implementation of new ways of operating.

9

Procedures/Job Aids

The console operator and I were reviewing his actions in response to a refinery wide power failure. Since he had three units under his span of control, he had three three-ring binders of emergency procedures in front of him. I asked him if he was supposed to do all the steps in one, then all the steps in the other, and so on. Or, was he to do the first step in each and then the second step in each? He responded with a smile and a shrug, "Your guess is as good as mine." Unfortunately, that is about the response I expected.

Various incidents in process plants over the past 30 years have resulted in an increased emphasis on the use of procedures. While procedures have always existed, in the past they were thought of as suggestions, more of a set of guidelines than a code. Of course, when operators take the incorrect action, one of the suggestions is to document the correct action in a procedure and require that it be followed. However, this assumes that the procedures are "follow-able," which is often not the case. Related to, and interacting with procedures are job aids, various materials that can help an operator in the performance of various tasks. This chapter examines emergency procedures in depth, due to their importance, and job aids cursorily due to their limited, but high potential use.

9.1   UPSET ANALYSIS

What amazes me about emergency procedures is, like much of what we have discussed, how little thought and analysis goes into their development.

Numerous times I have asked operators for their response to a specific event only to be told, "That can't happen, we don't have a procedure for it." If only it were that easy. Or, I find out that the procedure is written for a particular set of assumptions that are not articulated, such as whether the power failure is plant wide or isolated to a particular unit. At one plant this had significant repercussions. If the tankfarm does not lose power like the unit, then the charge pump had to be blocked in or a rupture disc would blow, leading to a hydrocarbon spill, release, and potential fire. I had one unit in which I was told that a steam failure would be the most severe event. When I got to the control room, the operators told me it couldn't happen. I thought I had been set-up to look like a fool (I don't need much help there), so I began a discussion with the operators,

> Me: "Don't you get steam?"
> Them: "Oh yes, it's absolutely critical to our process."
> Me: "So you could have a loss of steam?"
> Them: "No, it's too important."
> Me: "If the steam comes in, then it could not come in."
> Them: "No, we are last on the steam shed list."
> Me: "But if no steam is being produced, you would eventually run out?"
> Them: "No."

It took a good 20 min to convince them that a steam failure could occur.

Before writing the procedures, the potential upsets that require procedures need to be identified. This is done for two key reasons. First, such an analysis should ensure that all the procedures that are needed are written. Second, the analysis will likely reveal commonalties across events. You don't have to walk through too many emergency procedures before you start to see the same steps repeating. The repetition has two sources. First, a major event (such as loss of charge pump) is often just a subset of an even larger event (such as power or steam loss). The loss of any motor-driven piece of equipment is a subset of a power failure. Second, emergency procedures generally result in shutting down the unit, so the procedures converge on those tasks associated with a shutdown.   shows the interactions of upsets for a major process plant. A power loss resulted in, among other things, pump loss, compressor loss, and cooling water loss, leading to steam loss. Each of those events is a subset of the power loss event. Not only did it show that all upsets were a subset of a power failure, but pointed out a lack of robustness in plant design, such that a power failure would also be a steam failure due to a loss of cooling water.

For each upset, it is important to determine its impact and the objectives for each position. Too often, procedures are written for the unit, but procedures are for the benefit of the individual, the person performing a specific job. By defining the high-level impact and objectives, the key items in the procedure

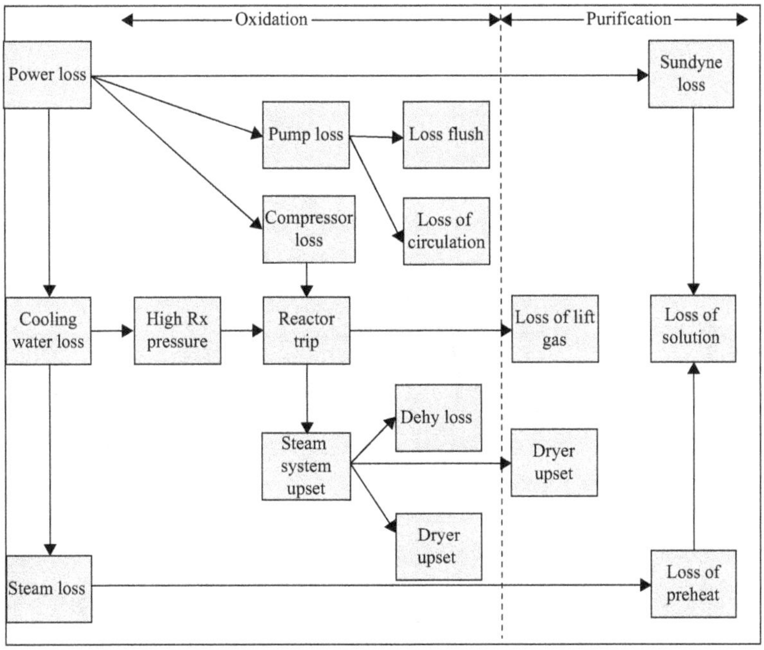

Figure 9.1.   Upset Interactions.

that need to be covered are addressed. If the operator is not impacted, then those personnel who will likely need their assistance should be specified. At a minimum, loss of major utilities (steam, power, air) and loss of major rotating equipment (compressors, charge pumps) should be evaluated. Determine what the plant will do in the calm setting of writing the procedure, not in the middle of the upset. What are the general objectives for the operators, for example, go on circulation or pull feed, shutdown heaters or go to minimum firing? Figure 9.2 shows an example of this high-level analysis of the impact of different upsets on different positions.

I want to caution you about the notion that emergency procedures are the key to high reliability operations. As with most engineering, emergency procedures are written under the double-jeopardy principle: two independent failures are not considered in the analysis. However, most major plant accidents I have examined involved just that, two independent failures. James Reason uses the Swiss cheese model of failures in layers of protection to analogize that it is possible for all the probabilities to align, such that the multiple holes in a block of Swiss cheese results in a hole all the way through, the analogy for the path of failures [54]. You will not have a procedure for that.

Overproceduralization is an area of concern for me. Where procedures were once just a memory aid for the user, they are increasingly becoming standards of behavior that are to be enforced. Not surprisingly, many

	Water operator	Heater operator	Frac operator	Compressor operator	Hydro operator
Cold box tube rupture	Steam production & BFW demand will drop as heaters/feed are cut	Feed will need to be ramped down, but not so fast as to coke off tubes	Cold box will need to be isolated, blocking all in/out streams	Compressors will need to be stopped/secured with shutdown of cold box	Will lose feed to hydro
Power failure	Get heaters down. Everything is done in manual	Manually divert to coke trap and get heaters down	Lost pumps-block everything in. Go to off-test; get Rx/settler stabilized. Won't have to isolate all cold box vlvs.	EMV's won't work-will have to be done manually and are large.	Will lose H2 because PSA will trip. Will see temp excursions and comp will trip. Send everything to off-test. Manually operate vlvs to manage levels.
Steam failure	Will lose turbines, switch to electric. Will have to S/D furnaces if there aren't enough people-possibly take down one heater and slow the rest (best case).	Valves are in manual	Will go cold. Won't have to isolate all cold box vlvs. Go to off-test; get Rx/settler stabilized.	Will have a little time to secure unit.	Put to off-test and manage levels. Use H2 to cool.
Instrument air loss	Concerned with polishers—don't let slam shut and rob of water. Bypass DMV's.	BFW DMV failed open to steam drum-need to manually block in so not to puke over to superhtr	Won't have to isolate all cold box vlvs. Go to off-test; get Rx/settler stabilized.	Air operated valves will instantly trip	Put to off-test. Will lose PSA.

Figure 9.2.   Example of the Impact of Different Upsets on Different Positions.

companies are developing sets of memory aids that are not called procedures, but would have been previously, because changing a procedure has become too arduous a task in the organization, often involving the legal department (who know a lot about operating a process). "Procedures" are becoming ossified as they are turned into legal tools to reduced liability. This increases the rigidity of the system [55], the opposite of resilience that is needed in many complex systems.

9.2 ORGANIZATION

As I alluded to earlier, one of the largest mistakes in procedures is their focus: the unit. Procedures tend to be written, apparently, for the hardware, with a procedure for each type of failure the unit could experience. But units are not jobs or positions. A person has a position, which may entail multiple units. The procedure should be written for the position, the person. So specify what actions are needed on all the units for which they are responsible. Large units with multiple operators have a slightly different variation of this problem. While the unit focus ensures that every operator's actions are covered, the procedures are often so long and cumbersome as a result that finding and extracting the steps an individual operator is to perform is often difficult. Finally, how do you handle the console operator with multiple units whose field operators may have only that unit as their responsibility?

The procedure question was the focus of a study commissioned by the Center for Operator Performance. A pilot effort along with industry experience indicated that modularizing the information in the procedures had significant potential to improve both presentation and ease of updating. Sandeep Purao of Penn State University undertook an effort to create a tool that could read existing procedures and apply heuristics to parse them into parts of speech [56]. Related sets of actions could be chunked into a module, with the operator associated with the action linked to the module. The result is a database of procedure modules that can be combined for a given upset by the unit (the traditional, although wrong way) or by the position. While it was suspected that there was significant duplication in the procedure content, the project revealed that less than 10% of the information was unique. What were four pages of instructions had been rewritten and recombined into over 40 pages of procedures for a single unit. The 10% unique information was 171 sets of actions that could be grouped into 23 modules, as in Figure 9.3.

Several other benefits were noted from the modularized procedure approach. One of the main benefits was once a module is modified or revised,

#	Chunk	Steps
1	Clear exchanger tubes	2
2	Shutdown unit	2
3	Isolate frac	4
4	Make notifications	4
5	Total reflux	1
6	Secure heaters	3
7	Trip charge heater	2
8	Total recycle	3
9	Troubleshoot cause	3
10	Line-up to offspec	4
11	Secure amine and H2	4
12	Bottle in 900 unit	13
13	Charge heater ESD	10
14	Conserve H2	32
15	Depressure unit	7
16	Event follow-up	15
17	Frac reboil heater ESD	9
18	····formerly initiate SD	1
19	Isolate feed rundown	8
20	Low H2 operation	8
21	Prevent runaway	14
22	Secure 900 unit	18
23	Shutdown charge	4
	Total	171

Figure 9.3.   Example of Chunking of Procedures.

it is modified everywhere it is used. This will minimize the potential for conflicts between different procedures. Because the modularization has such a significant reduction in volume of information, the process of reviewing and updating procedures is not so daunting. Finally, because it becomes clear that operator actions in an upset are a far smaller set than was previously thought, training programs can be built around the modules to further enhance operator performance.

9.3 FORMATTING

Procedures are just a type of display, a means of transferring information, so the formatting of the information follows identification of the content. There are some general guidelines for formatting procedure information.

As obvious as this may sound, steps should be in chronological order. Surprisingly, this is not always done. Even though two steps relate to the same piece of equipment (e.g., block fuel gas and steam heaters), if the steps are not to be done at the same time (e.g., heaters will be steamed some time prior to start-up), then do not place the steps together.

If multiple operators are needed, the steps should indicate which operators are to perform which actions. For example, if two valves in separate parts of the unit need to be isolated to go on circulation, indicate that the action should be performed by different operators. Too often, it is assumed that some operator will arrive to help complete a two-person task, but if it is not explicitly stated in the procedure, how will the operator who is supposed to assist know it is a requirement?

The actual steps in the procedure should be following the following rules: (Note: All "poor" examples are taken from actual Emergency Procedures)

- Provide a clear and meaningful title
  - *Appropriate*: Heater 29 Tube Rupture.
  - *Not appropriate*: DHT Shutdown Procedure Heater 29 Tube Rupture Emergency Procedure
- Be concise, the person reading the procedure is a trained operator
  - *Appropriate*: Maintain unit pressure at 250 PSIG. Purge reactor with hydrogen until RIT is less than 500°.
  - *Not appropriate*: Maintain unit pressure at 250 PSIG by taking hydrogen via 108 G and venting until reactor temperature is below 500°.
- Are warnings and/or cautions placed before the step to which they refer?
  - *Appropriate*: Caution: Ensure converter feed is present before starting pump CC-635. Start pump CC-635.
  - *Not appropriate*: Start pump CC-635. Do not start if converter feed is tripped.
- Avoid ambiguities, words that are subject to interpretation (high, low, fast, slow, enough, after a while, etc.) Provide specific values or range of values.
  - *Appropriate*: Adjust 5-PRC-141 setpoint to control suction pressure at 8 PSIG.
  - *Not appropriate*: Adjust 5-PRC-141 setpoint to maintain suction pressure at desired level.
- Highlight decision points.
  - *Appropriate*: When
    - reactor temperature is at or below 500°F, or
    - separator temp is above 200°F, or
    - compressor valve temp is above 300°F,
    - Shutdown all compressors
  - *Not appropriate*: Continue circulating hydrogen until reactor temperature is below 500°F, or separator temperature is above 200°F, or compressor valve temperature is above 300°F. Shutdown compressors 503 and 504.

- Identify whether the task is to be done inside or out.
  - ▫ *Appropriate*:
    - – Open dampers to heater 29 (inside or remote).
    - – Open dampers to heater 30 (outside or local).
  - ▫ *Not appropriate*: Open dampers to heaters 29 and 30.
- Is it clear what happens if a condition is not met?
  - ▫ *Appropriate*: Close pneumatic feed valve if possible. If not, close discharge block.
  - ▫ *Not appropriate*: Close pneumatic feed valve if possible.
- Do not place multiple tasks in a step.
  - ▫ *Appropriate*:
    - – Close individual fuel and burner block valves
    - – Shutdown air fans
    - – Shutdown feed and product inhibitor injection
    - – Unload compressors
  - ▫ *Not appropriate*: As time permits, close individual fuel and pilot burner block valves. Shutdown air fans, and feed and product inhibitor injection. Unload compressors.
- The alpha-numeric and descriptive title should be given for both equipment and instruments
  - ▫ Example:
    - – Slurry Recycle Pump (CC-604)
    - – C Column Bottoms Temp (3-TCI-210)
- If a plant condition can only be maintained for a short period, indicate the limiting parameters for operating in that condition. The parameters can be time, temperature, pressures, flows, and so on.
  - ▫ *Appropriate*: (loss of feed) Circulate rich and lean DEA until pressure = 220 PSIG.
  - ▫ *Not appropriate*: (loss of feed) Circulate rich and lean DEA.

9.4   JOB AIDS

Related to procedures are job aids. This is additional information for a specific task that may be embodied in a text document or posted at the task site. Job aids are not frequently used, but research indicates they should be used more often. The Center for Operator Performance provided a research topic for a senior design class at Wright State University. The students evaluated performance for conducting tank line-ups using different types of job aids (Figure 9.4). A tank manifold from a refinery tank farm was duplicated using home plumbing pipe and valves from a hardware store. Student volunteers underwent training on the manifold and performed three tasks considered easy, moderate, and difficult.

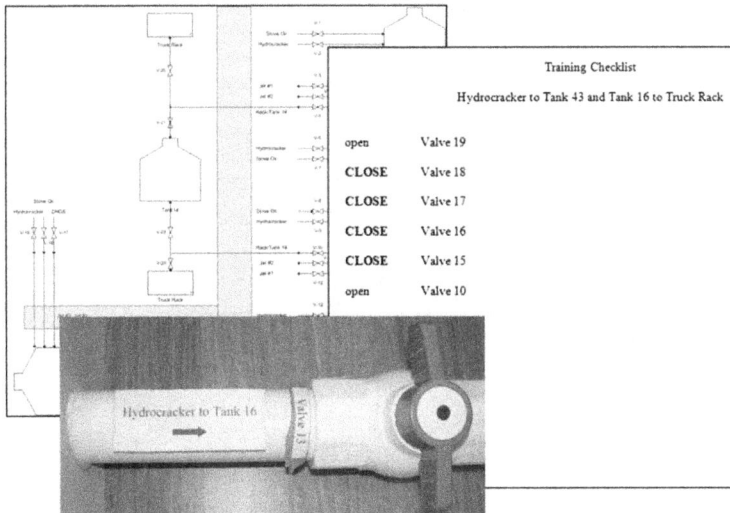

Figure 9.4. Tank Line-Ups Using Different Types of Job Aids.

A week later, they performed three more tasks. There were five conditions under which the tasks were performed:

1. Valves were labeled, but lines not marked.
2. Lines were labeled as to their destination.
3. Diagram of entire manifold provided.
4. Check list of required changes provided.
5. Combinations of above.

The researchers found that accuracy was significantly better when using any of the job aids over no job aids ( ). No jobs aids result in a slightly better than chance performance. On the opposite side of the performance continuum was use of a checklist with line demarcations, which resulted in perfect performance regardless of whether they also had the diagram. Strangely, the diagram with a checklist resulted in worse performance than just the diagram alone. This appeared to be a case where the additional information created confusion, thereby degrading performance. The impact of too much information degrading performance is not unique to this situation.

I have found operators repeatedly confounded by performing line-ups. When we do upset walk-throughs, lining-up to off-spec tankage is clearly confusing at many manifolds. I was with an operator that was attempting to send off-spec material to an off-spec silo. They did the line-up incorrectly, but prevented contamination of good product by chance; the silo that received the material happened to be empty at the time. Given the difference in the error

Accuracy

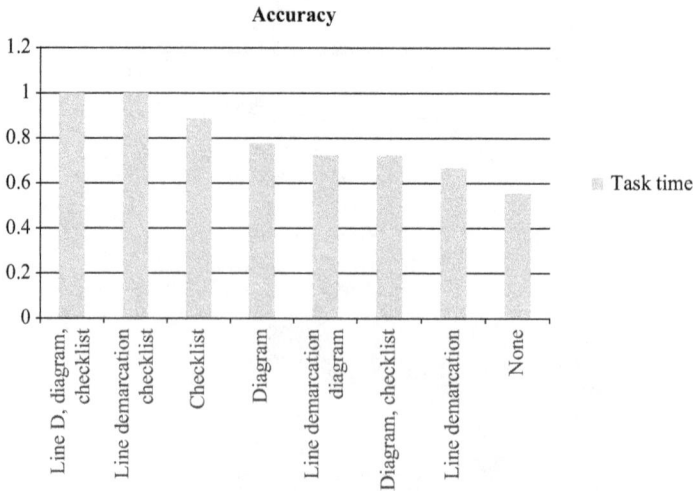

Figure 9.5. Accuracy of Using Various Job Aids to Perform a Line-Up.

probability from the student study, it would seem that a checklist and demarcation should be routine for any line-up changes. However, it rarely is.

9.5 CONCLUSION

Operator performance can be improved by providing a list of tasks or instructions, not relying on operator memory. However, maximizing the benefit of such a list requires that it be the product of careful analysis and design, not a stream-of-conscious dump of an experienced operator's memory.

10

Conclusion

I was in a new central control room of a refinery that had large monitors for an overview display at each console. "I can't read what's on there, can you?" I asked several of the company personnel standing with me. It wasn't that I didn't understand what was on the screen; the contrast was so poor that I couldn't even make out what was written (white cow in a snowstorm effect). They all replied that they couldn't read it either. "Don't you think you should be able to read it?" I challenged them. "Oh, that's what the human factors experts say it should be like" was the response.

I tell this story to warn against taking the advice of human factors exerts (or any expert) without critical analysis. Hopefully, this book has provided you with some ammunition to examine and assess changes designed to enhance operator performance. In an effort to highlight how this might be done, I want to examine the official report on one of the pivotal human factors incidents in the process industries—the release and explosion at the then British Petroleum's (BP) Texas City Refinery. I want to emphasize that any criticism of the findings is not intended to diminish the tragedy that occurred. However, inaccurate findings can be more damaging than none at all, as the former can result in changes that have unintended consequences, some of which may actually reduce safe operation rather than improve it.

Most industry professionals know the rough details of the event. A start-up of the raffinate splitter tower in the isomerization plant (ISOM) unit on March 23, 2005, resulted in the overfilling of a tower and subsequent release to a blowdown drum. Vapors from the drum ignited, with the fire and explosion causing the deaths of 15 workers, most of whom were in trailers adjacent to the unit. An investigation of the event was conducted by the Chemical Safety Board (CSB), which detailed several human factors as contributing to it. Let's look at their assessment.

The CSB listed nine underlying latent conditions that contributed to the unsafe start-up [57]:

1. A work environment that encouraged operations personnel to deviate from procedure.
2. Lack of a BP policy or emphasis on effective communication for shift change and hazardous operations (such as unit start-up).
3. Malfunctioning instrumentation that did not alert operators to the actual conditions of the unit.
4. A poorly designed computerized control system that hindered the ability of operations personnel to determine if the tower was overfilling.
5. Ineffective supervisory oversight and technical assistance during unit start-up.
6. Insufficient staffing to handle board operator workload during the high-risk time of unit start-up.
7. Lack of a human fatigue prevention policy.
8. Inadequate operator training for abnormal and start-up conditions.
9. Failure to establish effective safe operating limits.

I am not in agreement with some of these findings and having read this book, I feel you would be thinking the same. Obviously, I don't have access to all their data or analyses, and I have not put the time into the analyses that they have. However, their published results bring into question the conclusions they have drawn on the human performance aspects of the event. In some cases, I do not disagree with the conclusion, but it was not clear from the report that it was supported by the data. Let's look at their findings in the context of human factors variables.

10.1 SYSTEM DEMANDS

Much of the report and associated findings relate to the unit being in a "start-up." They assume, to a degree correctly, that a start-up places unique demands on the operator and warrants particular requirements, as it is not a normal event. However, start-up is not a clearly defined term or mode. The CSB report repeatedly uses the term as it would be used with the start-up of a large integrated unit like a fluid catalytic cracker (FCC) or an integrated set of units like a hydrotreater and reformer. However, this was a tower, not an entire unit.

Start-up is a continuum, from start-up of the FCC to start-up of a pump. This was certainly more than the start-up of a single pump, but it also was less than the start-up of a large unit.

10.2 WORKLOAD AND STAFFING

CSB finding: Insufficient staffing to handle board operator workload during the high-risk time of unit start-up.
My view: Not supported by the evidence provided.

The CSB references a BP analysis that says control of all units in the area (NDU, AU2, and ISOM) requires 10.5 h of a 12 h shift on a normal basis. This translates into 88% of the operator's time, absurdly high. I worked with an engineer who firmly believed that anything that came from an equation must be right, but this answer should appear unreasonable to anyone who has spent some time in a control room. The very nature of the console job, that is, performing large numbers of intermittent short duration tasks, precludes console operators from being able to spend anything close to 88% of their time on job-related activities. It should be remembered that there is a trade-off in time on task and the number of tasks. Console operators have far less time on task than field operators due to a far greater number of tasks. Our studies of over 300 console operators show the average console operator to spend only about 50% of their time on job-related tasks. The staffing need was said to be exacerbated by being in a "start-up," again confounding the different meanings of "start-up." I was with an operator who was shutting down a hydrotreater while we listened to the beginning of Operation Desert Storm on the radio. He didn't miss a beat.

10.3 OPERATOR–PROCESS INTERFACE

CSB finding: Malfunctioning instrumentation that did not alert operators to the actual conditions of the unit.
My view: I totally agree.

As far as I am concerned, this is *the* issue in the accident. Three instruments had failed. The operator had indication (though erroneous) that the level was decreasing. The fact that the redundant level switch alarm did not actuate supplied confirmatory evidence. The level switch in the blowdown drum did not actuate until the explosion (I would speculate that the blast shaking the switch was the cause). How many instruments should the operator have distrusted and continued to search for the real answer? The CSB implies that even with two data points indicating adequate level, the operator should have searched for more. How long do you search for something that might not exist? This strikes me as Monday morning quarterbacking, implying that they should have searched until the real problem was found despite evidence saying it was not a problem. One item not in the analysis was the status of the level in the overhead accumulator. That would seem to be the more proximate indicator of liquid carryover.

> *CSB finding*: A poorly designed computerized control system that hindered the ability of operations personnel to determine if the tower was overfilling.
> *My view*: Could have been better, but they saw a level and it was decreasing.

I again do not doubt the less than stellar quality of the interface at Texas City, but I disagree with some of the conclusions of the CSB. The CSB repeatedly mentions doing material balance calculations and having the procedures aid with performing them. While that is a potential option, it would be far easier to show trends of inlet and outlet flows. As the inlet flow changes, one or more of the outlet flows should change in the same direction. I have attempted to create material balance indicators and it is not as easy at is sounds. Flashing of material across the flow valve and transmitter can create less than reliable data.

In describing the problems of the interface ("The Board Operator—far removed from the physical location of the process unit undergoing startup—depended on the system to provide him with crucial process information, which requires a well-designed control board."), the CSB implies that the distance of the control room to the unit was an issue. The distance was not explicitly listed as a contributing factor, but the fact that it was raised at all makes it appear that it might be. It shouldn't have been a factor. Control of units is successfully done from a wide range of distances, whether in terms of miles, in the case of refineries, or across states, in the case of pipelines.

10.4   SELECTION AND TRAINING

> *CSB finding*: Inadequate operator training for abnormal and start-up conditions.
> *My view*: Of course, but not related to the incident.

I have no doubt that the operator training program at Texas City was lacking, as almost all process plant training programs are. However, the lack of specific training on tower overfilling is not the problem. If you asked any operator what would happen if you put flow into a vessel and didn't allow any out, a very high percentage would say it would fill and come out wherever possible. This would be akin to asking participants in a gun safety course to specify what would happen if you put a loaded gun to your head and pulled the trigger. Failure to "train" on the obvious is not the issue.

The CSB criticizes BP for not using training simulators, as in the aerospace and nuclear power models. However, my experience in process plants is that simulators are used for start-up and then quickly fall into disuse. It would not be wrong to use simulators, but the training needs in process plants are different than aerospace and nuclear power contexts. Where aerospace and nuclear power have a few basic designs with multiple productions, process plants have a wide variety of units and unit designs assembled in ways unique to each plant. As such, a different type of training in simulation is likely to be needed.

A further criticism of BP was the cut in the training department staff from 28 to 8. This was inferred to be indicative of a lack of concern for training. It may well have been a concern for not wasting money. There is no evidence that the greater number of staff produced better operators and operations. If it did, it is unlikely that they would have been cut. If there was a quantifiable link between training and plant performance, then there would be an easy way to determine the value of additional training resources. A general lack of knowledge on how to best train operators still plagues most companies.

10.5 JOB AND ORGANIZATION DESIGN

> *CSB finding*: Lack of a BP policy or emphasis on effective communication for shift change and hazardous operations (such as unit startup).
> *My view*: Of course, but not likely to have contributed to the incident.

Like procedures, good communication at shift change is important. I certainly have seen my share of "it's Cadillac-ing" or shift change that occurs at the turnstile. However, to attribute the Texas City incident to bad shift change practices is what I wouldn't agree with.

Most operators at shift change take a "trust, but verify" approach or attitude with regard to the information they are given. After relief is made, the operator is likely to do their most thorough check of the process. This includes checking on what they were told were issues, but also checking to see what might not have been said. I would argue that the fact that shift change occurred, regardless of the turnover quality, increased the potential to identify the problem, not decrease it.

CSB finding: Lack of a human fatigue prevention policy.
My view: Fatigue was likely a factor, but not proven by evidence provided.

Given the continuous days that the operators worked, I have no doubt that they were fatigued. However, the logic used in the CSB to state that fatigue was a contributing factor is weak. Their analysis can be captured in a syllogism:

- Fatigue causes poor judgment.
- The operators exhibited poor judgment.
- Therefore, the operators were fatigued.

That syllogism would work if it started with "fatigue causes all poor judgments." But it doesn't because there are many reasons why poor judgments occur. Hence, the fatigue discussion is a circular argument: The operators must have been fatigued because they exhibited poor judgment, which can be caused by fatigue. It would have been sufficient to list the continuous days worked and state that fatigue may have contributed to the result.

CSB finding: Failure to establish effective safe operating limits.
My view: Of course, but not likely to have contributed to the incident.

What does it mean to be safe? While that may sound silly, it turns out not to be so easy to determine when discussing safe operating limits. I have been involved in a variety of efforts to determine those limits, and it has generated heated discussions. Is it unsafe to lose cooling to an exothermic reactor? What if the reactor is down? Therefore, safety might be relative. It would seem that setting a safe limit for a pressure vessel should be easy. However, do you set it at the relief valve setting (typically 90% of the vessel rating), the vessel limit, or below the relief valve setting? Does it matter what material is in the vessel (hydrocarbon, steam, hydrogen sulfide)? What about temperature on metallurgy? The limits are in temperature for a given pressure, but the limit has to be on a single variable. Generally, exceeding the metallurgical limits will reduce the life of the metal but not result in the immediate catastrophic failure that would occur with mixing hydrocarbon and hot air. A set of safe operating limits is not needed for an operator to know that filling an entire column with hydrocarbon liquid is not a good idea.

> *CSB finding*: Ineffective supervisory oversight and technical assistance during unit start-up.
> *My view*: The term "start-up" can have multiple meanings and creates the logical fallacy of equivocation.

This finding is a product of the ambiguous nature of "start-up." Obviously the imposition of start-up administrative controls will vary with where on this continuum the start-up lies. However, the report generally treats the start-up of a single column at Texas City as though it were an FCC. As such, they set expectations for supervision and staffing that I think the start-up of a column would not typically warrant. They equivocate in defining the requirements for "start-up" as those of an entire unit and they use "start-up" to apply to a single column.

I had the opportunity to observe an FCC start-up where the CSB findings were taken seriously. First, it required significant waivers for us to even observe. The start-up itself was an eye opener in that everyone was jammed into the control room, such as the extra operators typical of a large unit start-up and also all the process engineers with laptops humming, and the unit manager. When I asked why he was there, I was told he is needed in case high-level decisions have to be made. Had anything gone wrong I would have dinged them for creating a chaotic work environment, but they were doing what they thought was needed post–Texas City.

10.6 PROCEDURES AND JOB AIDS

> *CSB finding*: A work environment that encouraged operations personnel to deviate from procedure.
> *My view*: Procedures in the industry were never intended to be followed verbatim.

I was shocked when I heard that "not following procedures" was listed as a causal factor in the upset. Not because I thought they would follow procedures, but I thought everyone in the business knew most procedures are unfollowable. One location had seven pages of preliminary information, including a statement that hardhat and safety glasses were to be worn when following the procedure (which was stating the obvious, given that nothing is to be done without hardhat and safety glasses), before any useful instructions were given. One of the best procedures I have encountered were checklists posted at the door of the control room with key tasks for different upsets. It is my understanding that they no longer exist because it can't be proved that they are the latest procedure. A good tool has been eliminated because it might not be perfect. Never let the perfect become the enemy of the good.

Certainly good procedures are needed, but they do not exist in the process control industry yet. Regardless, there will always be a tension between accuracy and the ability to follow them. As all procedures are written with a given set of assumptions (initial conditions, nature of upset), the more specific the procedure, the more likely that it will not directly apply to a given upset [58]. A Catch-22 situation is created where either the procedure is followed, resulting in inappropriate actions, or it is not followed to the letter, thereby raising the specter of "violating" the procedure.

10.7 CONCLUSION

The BP Texas City instrument was clearly preventable. The fact that it wasn't prevented by the operator monitoring and controlling the process is a tragedy. At the heart of the problem is erroneous data on the level in the column that was being supplied to the operator. However, systems need to be robust enough that bad data does not result in the loss of life and type of damage seen at Texas City. I suspect the conflating issue was a failure to note the system was not behaving as expected, a key feature of expertise seen earlier. Without more analysis, I cannot determine whether that inability to detect abnormal was a product of interface, training, or expectations. Hopefully, you will have concluded by now that it was likely a product of all of them and more.

Human factors experts are human. We make mistakes. I hope that having read this book you now have a greater appreciation and insight into the human factors that influence operator performance. Equally important, I feel that you are able to critically assess human factors recommendations that are provided to you or the industry. If you find the topic interesting, I suggest that you may want to get involved in the Center for Operator Performance (www.operatorperformance.org), an industry–academia collaboration researching ways to enhance plant safety through better operator–process system design.

References

[1]. Meister, D. "The Formal History of HFE," In *The History of Human Factors and Ergonomics*. Mahwah, NJ: Lawrence Erlbaum Associates, 1999.

[2]. Sanders, M.S.; and E.J. McCormick. *Human Factors in Engineering and Design*, 370–371. 6th ed. New York, NY: McGraw-Hill, 1987.

[3]. Sheridan, T.D.; and W.R. Ferrell. *Man-Machine Systems: Information, Control and Decision Models of Human Performance*. Cambridge, MA: MIT Press, 1975.

[4]. Stroop, J.R. "Studies of Interference in Serial Verbal Reaction." *Journal of Experimental Psychology* 18, no. 6 (December 1935), pp. 643–662.

[5]. Miller, G.A. "The Magical Number Seven, Plus or Minus Two: Some Limits on Our Capacity to Process Information." *Psychological Review* 63, no. 2 (March 1956), pp. 81–97.

[6]. Klein, G. *Sources of Power: How People Make Decisions*. Cambridge, MA: MIT Press, 1998.

[7]. Wickens, C.D. *Engineering Psychology and Human Performance*, 345–347. Harper Collins, New York, 1992.

[8]. Goodstein, L.P.; H.B. Andersen; and S.E. Olsen. *Tasks, Errors and Mental Models*. London, UK: Taylor and Francis, 1988.

[9]. International Society for Automation. *Management of Alarm Systems for the Process Industries* ISA 18.2. Research Triangle Park, NC: ISA, 2009.

[10]. Uhack, G.; and C. Harvey. "Empirically Evaluating and Developing Alarm Rate Standards for Liquid Pipeline Control Room Operators," *3rd International Conference on Applied Human Factors and Ergonomics*. Miami, FL, July 17–20, 2010.

[11]. Sorkin, R.D.; B.H. Kantowitz; and S.C. Kantowitz. "Likelihood Alarm Displays," *Human Factors* 30, no. 4 (August 1988), pp. 445–459.

[12]. Cummings, M.L.; and S. Guerlain. "Developing Operator Capacity Estimates for Supervisory Control of Autonomous Vehicles," *Human Factors* 49, no. 1 (February 2007), pp. 1–15.

[13]. Kahneman, D. *Attention and Effort.* Englewood Cliffs, NJ: Prentice-Hall, 1973.

[14]. Wickens, C.D. *Engineering Psychology and Human Performance,* 393–396. New York, NY; Harper-Collins, 1992.

[15]. Schneider, W. "Training High Performance Skills: Fallacies and guidelines," *Human Factors* 27, no. 3 (June 1985), pp. 285–300.

[16]. Logan, G.D. "Automaticity, Resources, and Memory: Theoretical controversies and practical applications," *Human Factors* 30, no. 5 (October 1988), pp. 583–598.

[17]. O'Donnell, R.D.; and F.T. Eggemeir. "Workload Assessment Methodology." In *Handbook of Perception and Performance,* eds. K. Boff; L. Kaufman; and J. Thomas(Vol II). New York, NY; Wiley, 1986.

[18]. Koffskey, C.; L. H. Ikuma; and C. Harvey. Performance metrics for evaluating petro-chemical control room displays. *Proceedings of the Human Factors and Ergonomics Society 57th Annual Meeting,* September 30–October 4, 2013, San Diego, CA.

[19]. Wickens, C.D. *Engineering Psychology and Human Performance,* 413–415. New York, NY; Harper-Collins, 1992.

[20]. Young, M.; and N. Stanton. "Malleable Attentional Resources Theory: A New Explanation for the Effects of Mental Under Load on Performance," *Human Factors* 44, no. 3 (September 2002), pp. 365–375.

[21]. Klein, G. *Sources of Power: How People Make Decisions.* pp. 45–77. Cambridge, MA: MIT Press, 1999.

[22]. Klein, G. "Features of Team Coordination." In *New Trends in Cooperative Activities: Understanding System Dynamics in Complex Environments,* 68–95, eds. M. McNeese; E. Salas; and M. Endsley. Santa Monica, CA: Human Factors and Ergonomics Society, 2001.

[23]. Sellie, C.N. "Stopwatch Time Study." In *Maynard's Industrial Engineering Handbook,* ed. W. Hodson, 4.31–4.32. 4th ed. New York, NY; McGrawHill, 1992.

[24]. Norman, D. *Things That Make Us Smart,* 105. New York, NY; Basic Books, 1993.

[25]. Hawkins, J. *On Intelligence,* 44–45. New York, NY; Owl Books, 2004.

[26]. Wickens, C.D. "Information Processing, Decision Making, and Cognition," In *Handbook of Human Factors,* 91, ed. G. Salvendy. New York, NY; John Wiley & Sons, 1987.

[27]. Van Cott, H.; and R. Kincade (eds). *Human Engineering to Equipment Design,* 391–400. New York, NY; McGraw-Hill, 1972.

[28]. U.S. Department of Defense. *Design Criteria Standard, Human Engineering*, MI-STD-1472F. Washington, DC: US Department of Defense Design Criteria Standard, 1999.

[29]. McCormick, E.J.; and M.S. Sanders. *Human Factors in Engineering and Design*, 53. New York, NY; McGraw-Hill, 1987.

[30]. Green, M. "Color for Text and Graph Legibility," *In Visual Expert Human Factors*, 2007, www.visualexpert.com/FAQ/Part6/cfaqPart6.html, (accessed December 1, 2008).

[31]. Najjar, L. *Using Color Effectively*, 1990, www.lawerencenajjar.com/papers/using_color_effectively.html, (accessed December 1, 2008).

[32]. Hill, A. *Readability of Websites with Various Foreground/background Color Combinations, Font types and Word Styles*, 1997, www.laurenscharff.com/reserach/AHNCUR.html, (accessed December 1, 2008).

[33]. Tullis, T. "An Evaluation of Alphanumeric, Graphic, and Color Information Displays." *Human factors* 23, no 5 (October, 1981), pp. 541–550.

[34]. Post, D.L. "Color and Human-Computer Interaction." In *Handbook of Human Computer Interaction*, eds. M. Helander, T. Landuaer, and P. Prabhu, 608. Amsterdam, The Netherland: Elsevier Science, 1997.

[35]. Ganriel-Petit, P. "Applying Color Theory to Digital Displays." In *UX Matters*, 2007, www.uxmatters.com/MT/archives/000163.php, (accessed December 1, 2008).

[36]. Massimo, G.; N. Stucchi; D. Zayagno; and B. Marno. "On the Portability of Computer Generate Presentations: The Effects of text Background Color Combinations on Text Legibility." *Human Factors* 50, no. 5 (October 2008), pp. 821–833.

[37]. Reynolds, L. "Colour for Air traffic Control Displays." *Displays* 15, no. 4, (October 1994), pp. 215–225.

[38]. Morton, J. *Color and Vision Matters, Color Matters*, 2008, www.colormatters.com/optics.html, (accessed December 1, 2008).

[39]. Gatlin, C. "Secrets of Web Color Revealed." In *York Structural Biology Laboratory, University of York*, 2001, www.ysbl.york.ac.uk/~mgwt/KKwebcourse/colourscience, (accessed December 1, 2008).

[40]. Smallman, H.S.; and M. St. John. "Naïve Realism: Misplaced Faith in Realistic Displays." *Ergonomics in Design* 13, no. 3 (June 2005), pp. 14–19.

[41]. Noe, A. *Action in Perception*. 67–70. Cambridge, MA: MIT Press, 2004.

[42]. Norman, D. *Things That Make Us Smart*. 47. New York , NY; Basic Books, 1993.

[43]. Klein, G. *Sources of Power: How People Make Decisions*. 287–289. Cambridge, MA: MIT Press, 1999.

[44]. Smallman, H.S.; and M. Hegarty. Expertise, spatial ability, and intuition in the use of complex visual displays. *Proceedings of the Human factors and Ergonomics Society*, 2007, pp. 200–204.

[45]. Endsley, M.; and J. Jones. "A Model of Inter- and Intrateam Situation Awareness: Implications for Design, Trainin, and Measurement." In *New Trends in Cooperative Activities: Understanding System Dynamics in Complex Environments*, eds. M. McNeese; E. Salas; and M. Endsley, 46–67. Santa Monica, CA: Human Factors and Ergonomics Society, 2001.

[46]. Sohn, M.; M. Douglass; and J.R. Anderson. "Characteristics of Fluent Skills in a Complex, Dynamic Problem-Solving Task." *Human factors* 47, no. 4 (Winter 2006), pp. 741–752.

[47]. Caro, P. "Aircraft Simulators and Pilot Training." *Human Factors* 15, (1973), pp. 502–509.

[48]. Schendel, J.D.; and J.D. Hagman. "Long-Term Retention of Motor Skills." In *Training for Performance: Principles of Applied Learning*, eds. J.E. Morrison, 53–92. New York, NY; John Wiley, 1991.

[49]. American Petroleum Institute. *Fatigue Risk Management Systems for Personnel in the Refining and Petrochemical Industries, Recommend Practice 755.* API Industrial Hygiene TF Workshop, Denver, Colorado, 2010.

[50]. Zalesny, M.; E. Salas; and C. Prince. "Conceptual and Measurement Issues in Coordination: Implications for Team Behavior and Performance." In *Research in Personnel and Human Resources Management*, ed. G. Ferris, 81–115, Vol. 13. Greenwich, CT: JAI Press, 1995.

[51]. Salas, E.; K. Wilson; C. Burke; D. Wrightman; and W. Howse. "A Checklist for Crew Resource Management." In *Ergonomics in Design*, 8–15, Vol. 14, No. 2. Santa Monica, CA: Human Factors & Ergonomics Society, 2006.

[52]. Salas, J.; K. Wilson; C. Burke; and D. Wightman. "Does Crew Resource Management Training Work? An Update, and Extension, and Some Critical Needs." *Human Factors* 48, no. 2 (Summer 2006), pp. 392–412.

[53]. Coonan, F.L.; and D.A. Strobhar. "Manage Plant Modernization Projects Humanistically." *Hydrocarbon Processing* 72, no. 5 (1993), pp. 55–62.

[54]. Reasons, J. *Managing the Risks of Organizational Accidents.* 9–13. Burlingtono, VT: Ashgate Publishing, 1997.

[55]. Fucks, I.; and Y. Dien. "No rule, no use? The effects of Over-Proceduralization." In *Trapping Safety Into Rules*, eds. C. Bieder; and M. Bourrier, 27–42. Burlingtono, VT: Ashgate Publishing, 2013.

[56]. He, J.; S. Purao; J. Becker; and D. Strobhar. "Service Extraction from Operator Procedures in the Process Industries." *Service Oriented Perspectives in Design Science Research: Lecture Notes in Computer Science* 6629 (2011), pp. 335–349.

[57]. U.S. Chemical and Hazard Investigation Board. *Investigation Report: Refinery Fire and Explosion*, BP Texas City, Report #2005-04-I-TX, March, 2007.

[58]. Reasons, J. *Managing the Risks of Organizational Accidents*. 51. Burlingtono, VT: Ashgate Publishing, 1997.

Index

THIS TITLE IS FROM OUR MANUFACTURING ENGINEERING
COLLECTION.OTHER TITLES OF INTEREST MIGHT BE...

Alarm Management for Process Control: A Best-Practice Guide for Design,
Implementation, and Use of Industrial Alarm Systems
By Douglas H. Rothenberg

Protecting Industrial Control Systems from Electronic Threats
By Joseph Weiss

Industrial Resource Utilization and Productivity: Understanding the Linkages
By Anil Mital, Arun Pennathur

Robust Control System Networks: How to Achieve Reliable Control After Stuxnet
By Ralph Langner

Raw and Finished Materials: A Concise Guide to Properties and Applications
By Brian Dureu

Quality Recognition & Prediction: Smarter Pattern Technology with the Mahalanobis-Taguchi System
By Shoichi Teshima, Yoshiko Hasegawa, Kazuo Tatebayashi

Going the Distance: Solids Level Measurement with Radar
By Tim Little

Plant IT Integrating Information Technology into Automated Manufacturing
By Dennis L. Brandl, Donald E. Brandl

Mastering Lean Six Sigma: Advanced Black Belt Concepts
By Salman Taghizadegan

Automated Weighing Technology: Process Solutions
By Ralph Closs

Automating Manufacturing Operations: The Penultimate Approach
By William M. Hawkins

Real Time Control of the Industrial Enterprise
By Peter Martin and Walter Boyes

Solids Level Measurement and Detection Handbook
By Joe Lewis

Announcing Digital Content Crafted by Librarians

Momentum Press offers digital content as authoritative treatments of advanced engineering topics, by leaders in their fields. Hosted on ebrary, MP provides practitioners, researchers, faculty and students in engineering, science and industry with innovative electronic content in sensors and controls engineering, advanced energy engineering, manufacturing, and materials science. **Momentum Press offers library-friendly terms:**

- perpetual access for a one-time fee
- no subscriptions or access fees required
- unlimited concurrent usage permitted
- downloadable PDFs provided
- free MARC records included
- free trials

The **Momentum Press** digital library is very affordable, with no obligation to buy in future years.

For more information, please visit **www.momentumpress.net/library** or to set up a trial in the US, please contact **mpsales@globalepress.com**.

www.ingramcontent.com/pod-product-compliance
Lightning Source LLC
Chambersburg PA
CBHW082008190326
41458CB00010B/3114